HOW TO
FORD & MERCURY
V8 ENGINES

By ROGER HUNTINGTON
MEMBER, SOCIETY OF AUTOMOTIVE ENGINEERS

Copyright 1951 by Floyd Clymer

4239 W. Ina Road, Suite 101
Tucson, AZ 85741
520-744-6110

History Revisited

Forty-seven years later I am pleased to reprint Roger Huntington's *How to Hop Up Ford & Mercury V8 Engines,* originally published in 1951 by Floyd Clymer in Los Angeles. Treasured by enthusiasts, by the 1990s these books had become almost impossible to find. Good copies were selling for $50 at swap meets.

I personally knew Roger Huntington and always appreciated his insightful articles on automobile design and especially on automotive performance.

Although he was a paraplegic, Roger never let that stop his search for automotive knowledge. You just never knew when you would see him in his wheelchair--at drag races, at press introductions at GM, Ford or Chrysler, out "test driving" as he rode along and got impressions of how a new car handled and performed. Then he would write about what he learned and how he felt about the car. His byline appeared in almost every automotive magazine of the day. Roger Huntington's name was synonymous with then-current knowledge about high performance.

I know you will enjoy reliving history as you turn the pages of this automotive performance classic.

Bill Fisher
September 15, 1997

Cover photo: Bob Tattersfield and Frank Baron assembled this display engine. It was photographed by Bill Fisher about 1949.

Preface

Okay, we might as well get one little matter over with right away! In any book of this kind, we've got to mention somewhere that "souping" — as we like to call it — is dirty business as far as stock car manufacturers are concerned.

What we're planning to do with the Ford V8 might be alarming to many Ford owners, and the torque we're plotting to put through that three-bearing lower end might keep Ford engineers awake nights! Many factory men won't get within a mile of any souping business.

And we can't really blame 'em. They've designed their little V8 to do a general transportation job. They've produced an engine that will get John Q. Public to work five days a week, winter and summer, take him 75 miles to Aunt Emma's for Thanksgiving dinner at an easy 60-mph cruise, and to do this day-to-day job for at least 50,000 miles without major repairs. Ford engineers have planned compression ratios, cam timings, pistons and they've laid out bearing areas and block sections to do just this job — and nothing more!

When we ask that same lower end to pump 200 hp without complaining, we're just asking for trouble. Souping isn't going to give us much of anything but a fast vehicle, remember that.

So let's give the manufacturer the benefit of the doubt on the deal. Let's not say the Ford engine is a stinker if the stock bearings throw in the sponge the 250th time up to 6000 rpm in low gear! We're subjecting that block to treatment it was never designed for. Laugh off your troubles and use your head to keep them from happening again. And if we happen to say something like: "The Jetto dual manifold adds 10 hp to the stock Ford V8 and gives smooth, crisp performance at all speeds" — you'll know we're not quoting the Ford engineering department!

In this book we're going to try to cover the full business of souping the Ford-Merc V8 engine. Ten years ago, this wouldn't have been too much of a job; only a relatively small amount of special equipment was available and the boys were satisfied with less hair-raising expedients.

But today, the V8 souping business has mushroomed into a snarling, dollar-gulping monster! Where they used to pour benzol in the tank at the race track, they're experimenting with explosives in alcohol today. Where they used to mill a cylinder head to 8:1 compression ratio, they're fitting overhead valves and 18:1 compression today. Where they used to be satisfied with heavier springs on the distributor breaker arms, they're worrying about magnetos on road cars today.

Oh yes, you've really got to squeeze to stay with the V8 crowd these days. It's a complicated and expensive business, with a thousand places to waste money. We can't hope to cover every shred of equipment down to the last bolt or every little souping refinement in this book. Every day, some guy is coming up with a new twist.

Neither are we going into minute detail on equipment installation and general machine work. It would take a 15-lb. volume and then some to do that. Machine work data is widely available from shop men or from other literature — and why should we spend 1,500 words telling you how to install a Mallory double-breaker distributor assembly when the full instructions are included with the equipment?

In this book, we just want to tell you what you can do, what you have to do with, how to plan the job, how to proceed, and how to get the most out of your "souping dollar." So let's away to the land of fantastic pressures, suicidal valve timings, and broken connecting rods!

ROGER HUNTINGTON

Announcement

In presenting this book, we are attempting to meet the increasing demands for specific information concerning the "souping" or hopping-up of stock Ford and Mercury V8 engines, which are very popular for use in racing, hot rod, and competition cars.

Following the same basic organization as he did in the book, "Souping the Stock Engine", author Huntington discusses effective theories on speed tuning, and pays special attention to cost economies — how to get the most for your investment in "hopping-up" these V8 engines.

That these speed theories are rewarding is well-evidenced by the performance of the Floyd Clymer *Motorbook Special*, a "super hot rod", which set an all-time high speed of 221.4795 mph in August of 1951 at the official time trials of the S.C.T.A. on the Bonneville Salt Flats. The car, designed by Bill Kenz and driven by Willie Young, of Denver, is equipped with *TWO* hopped-up Mercury engines. We were indeed happy to have sponsored this car, and to have been associated with the "fastest car in America" and the "fastest hot rod in the world!"

Although there is definite reference only to automotive installations, the information herein applies also to marine, aircraft, or any other situation where *increased power output* is desired.

In our opinion, this book is the most complete theoretical treatment of the subject ever published to date, and will become a standard text in the field of speed and power. We hope you enjoy it, and that you will derive real benefit from its contents.

Floyd Clymer
Publisher

List of Abbreviations.

HP	Horsepower (this refers to brake power delivered at the flywheel, not the British taxing HP based on bore and number of cylinders).
RPM	Revolutions per minute (usually referring to crankshaft speed).
MPH	Miles per hour.
MPG	Miles per gallon.
V. E.	Volumetric efficiency (explained in text).
%	Per cent.
lbs.	Pounds.
BTU	British Thermal Unit (explained in text).
c. c.	Cubic centimeters.
° F.	Degrees Fahrenheit.
sq. in.	Square inches.
cu. in.	Cubic inches.
BTC; ABC, etc.	Before top center; after bottom center, etc.

ACKNOWLEDGMENT

We desire to thank the following firms which assisted in supplying data on their products, photographs, and general information used in this book:

Ford Motor Company
Eddie Meyer Engineering Company
J. E. Engineering Corporation
Smith and Jones Company
Weber Tool Company
Navarro Racing Equipment Co.
Weiand Racing Equipment Co.
Ray Brown Automotive
Ed Iskenderian
Evans Speed Equipment Company
Harman and Collins, Inc.
Edelbrock Equipment Company
Witteman Company
Offenhauser Equipment Corp.
Crankshaft Company
Electronic Balancing Company
SpeedOmotive
Raceway Equipment Company
Stephens-Frenzel Company
Frenzel Engineering Company
Italmeccanica, Inc.
Besasie Engineering Company
Bell Auto Parts
Shell Auto Parts
Almquist Engineering & Mfg. Co.
Howard's Automotive
Newhouse Automotive Industries
So-Cal Speed Shop
Lee's Speed Shop
Ansen's Automotive Engineering
Bob Stelling
Speedsport
Eastern Auto Supply
Joe Hunt

Table of Contents

		PAGE
PREFACE		3
ANNOUNCEMENT		5
CHAPTER 1	WHAT'S IT ALL ABOUT?	9
	What Are We After—Why Soup—Get an Offy—More About Costs.	
CHAPTER 2	THE V8 FAMILY TREE	13
CHAPTER 3	PERFORMANCE FUNDAMENTALS	21
	Power and Torque—Performance Testing—The Torque Curve—B.M.E.P.—Two Flies in the Ointment—Fuel Consumption.	
CHAPTER 4	PLANNING THE JOB	31
	The Paths to Power—Limitations—Focusing on Costs—Cost vs. Results—Plan Ahead—Who Does the Work.	
CHAPTER 5	FITTING UP THE BLOCK	41
	Selecting a Block—Increasing the Bore—The Crankshaft — Pistons and Rings — Rods — Bearings — Lubrication — The Flywheel — Balancing — Cooling the V8 — Chroming.	
CHAPTER 6	CYLINDER HEADS	61
	Overheads vs. Side Valves — Compression Ratio — Reworking Stock Heads — Special Flat Heads—Overheads for the V8.	
CHAPTER 7	THE INDUCTION SYSTEM	77
	Fuels — Fundamentals of Gas Flow — The Carburetor — Manifolds — Porting and Relieving — Valves — Springs and Tappets — The Camshaft — The Exhaust System.	
CHAPTER 8	IGNITION	107
	Basic Fundamentals—Reworking Stock Ignition—Converted Dual Systems — Magnetos — Spark Plugs — Ignition Recommendations.	
CHAPTER 9	SUPERCHARGERS	125
	Why Bother — Supercharging Principles — The Centrifugal Type — The Roots Type — Choosing a Type — McCulloch — Frenzel — Besasie — Italmeccanica — Speedomotive.	
CHAPTER 10	WHAT'LL SHE DO?	141
	Layout Fundamentals—Estimating Peak H.P.—Working Out an Example — More About Souped Performance.	
CHAPTER 11	THE ALMIGHTY DOLLAR AGAIN	153
	"Souping" for the Short-Money Chap — If You Can Tear It Down — Cards on the Table.	

Fig. 1-1. The sale of speed and custom equipment to hot-rodders runs into millions of dollars a year. Speed shops now cover the U.S. Here's a view inside Shell Auto Parts, Los Angeles.

CHAPTER 1

WHAT'S IT ALL ABOUT?

WHEN Harry Miller was hired by the Ford dealer organization back in 1935 to design and build a fleet of semi-stock Ford racing cars — a skeptical automotive world watched the first serious attempt to soup up the famous "V8" engine.

The whole deal was a little ironical too. Here was Mr. Miller, world-famous designer of special race equipment, doing his best on a stock flat-head block to compete with his own overhead-cam engines! He did right well to squeeze 160 horses out of that little passenger-car plant, but his front-drive cars were always $3\frac{1}{2}$ seconds slower on lap time at Indianapolis than the 4-cylinder Millers. Anyway, even if the car wasn't a world-beater, we've got to give Miller the credit for taking those first faltering souping steps with the Ford V8.

And who would have predicted 15 years ago that backyard wrench-wrestlers would be pulling 250 hp from that basic block today? Sometimes progress moves in mysterious ways! But move it did, and today we find the Ford-Mercury V8 engine right at the head of the "souper's" list of raw material. Harry Miller was the pioneer of more than he ever realized!

WHAT ARE WE AFTER?

Before we get any further into this complicated souping business, perhaps it would be a good idea to pause right here and ask ourselves just what we're trying to do. What are we after? Certainly the American stock car engine — the Ford included — is no stinker when it comes to all-around performance, economy, price, wear, or anything else. Why are we hot to rework a piece of machinery that apparently is already doing a beautiful job?

Our Leadfoot Louie, the moron hot-rodder, pops up from behind a bench where he's "regrinding" his own cam with a file — and for once he makes a profound statement: "I'm souping up my V8 engine because I want more horsepower. I don't expect phenomenal gas mileage, she'll run rough and noisy, and I know the torque won't be worth sweat below 1500 revs, but I want 150 horses under that hood that I can tap when I want 'em."

Louie is right. Power is the big thing. But we'd like to ask Louie a couple of questions: If 150 hp is all you're after, a big straight-8 Packard engine would cost you less than your souped V8, and she'd run smoother, quieter, and the low-RPM performance would be a whole lot better. Why not get that? Or better still, you can easily pull 175 hp from a little 220 Offenhauser on pump gas — why not that?

Our Leadfoot Louie is dumbfounded, but shoots back, "That's fantastic. I've got a few little items like size, weight, and my pocketbook to think about, you know." And there you have it!

WHY SOUP?

Now we've hit smack at the real reasons for souping a stock engine. In other words, what we're after basically is not just a hot ENGINE, but a hot VEHICLE — whether it be a road car, a track racer, a boat, or an airplane. It's as simple as that.

You don't sink $500 in an engine to see it turn up 180 hp on Squeedunk Auto Parts' dynamometer. You do it to get a car that will all but crack your neck when you tromp on it — pick up from zero to 60 mph in nine seconds — or eat up Highway 10 at 170 ft. per second — or clock 122 mph at Bonneville or turn Gardena in 23 flat — or skim over Salton Sea at 80 mph! The vehicle is the thing, not the engine alone.

For this reason, an engine's size and weight are as important as its horses. We all know that the overall performance of a given vehicle, which includes acceleration and top speed, depends largely on its power, weight, and frontal area. To keep weight and frontal area to a minimum — that is, to get the maximum possible performance from a given horsepower — we've got to have a small, light, compact power plant. And souping doesn't appreciably alter an engine's size and weight.

That, in a nutshell, is why we engage in this madness. Obviously we get a lot more overall vehicle performance by souping a 90-hp, 400-lb. engine up to 160 hp than by fitting up a stock 160-hp job weighing 800 lbs. The Ford-Merc V8 engine weighs about 550 lbs. and can be quite readily souped to 180 hp in road trim; a fully stock engine in that power range would generally weigh twice this much (at least, before Chrysler and Cadillac began making overhead valve engines.) At any rate, the big reason behind our souping is to get a powerful, small engine — not just a powerful engine.

GET AN OFFY?

Then there's the question of using a full racing engine. Why not drop a little 400-lb. 220 Offy into your hot road chassis? Here's an easy 175 hp in a tiny, compact power plant, and at a weight you could never approach with a souped V8. And what a hot vehicle it would make, with that light, 4-throw crank!

Right here is where one of the nastiest gremlins we have to deal with in this souping business rears its ugly head — and that's the little item of costs! If we were all millionaires, speed equipment manufacturers and speed shops would fold overnight. Then we'd all buy ourselves an Offy — or maybe a 4½-litre 12-cylinder Ferrari, or some such $5,000 power plant— and all ride around in 300-hp cars. In other words, if you've got the dough, you can buy performance, and every other desirable engine feature, that will make any stock equipment look very ill.

But unfortunately, we're not all millionaires — hence this book. So if we expect to get upwards of 200 hp for an engine weight in the vicinity of 600 lbs., and for a total cost of say between $500 and $1,000 — we have no choice but to turn to the stock block. Ninety-nine percent of us can forget the Offys and Ferraris.

But there's one thing that should be kept in mind: If you plan to race for cash prizes with your equipment, definitely don't forget the Offy! Consider it carefully. For about twice what you'll pay for your souped stock engine, you can often pick up a fairly good used Offy that will out-pull it by a wide margin and weigh a couple hundred pounds less. Your chances of making your racing pay off will be a lot better.

We can lay no better emphasis to this statement than to cite the 1950 Indianapolis race; Bob Estes of Inglewood, Calif. entered a beautiful Merc-powered car, souped to the limit with Ardun overhead valves. But it was some 5 mph slower than the slowest qualifier in the top 33---Harry Miller's experience, 1950 style!

A lot of good Fords have trimmed a lot of poor Offys — but no Ford ever stayed ahead of a good Offy! It's too long and complicated a story to tell here, but the simple fact is that a stock block just can't compete on even terms with a special engine designed exclusively for racing. Too many guys have lost their shirts trying to prove it could! Remember this if you're going into racing.

Actually, the trend in American auto competition today is toward separating the special equipment from the semi-stock stuff. The big circuits (AAA, CSRA, etc.) encourage only the special double-overhead-cam engines, and you seldom see stock jobs running with these boys. On the other hand, the hot-rod and roadster organizations usually allow only stock blocks. There's not much mixed competition anymore, and that's the best for all concerned. At any rate, if the regulations don't bar double-overhead engines, we'd suggest that you think a long time before you jump into the battle with a stock block. The shirt you save may be your own!

MORE ABOUT COSTS

But even assuming you forget about special racing engines, your cost problems are far from over. Souping is definitely not a poor man's game, any way you look at it. You can't make a real move for much less than $50 and it's going to take a good $500 in your engine alone to give you a really hot vehicle. In fact, you can drop $1,000 without half trying!

Your average souping job on the Ford-Merc V8 will run around $250. Admittedly these costs might not break the average hot-rodder, but they're something to consider if you have a lot of other financial responsibilities. There's nothing so pitiful as the Leadfoot Louie with a family of four who can't really afford it, but sinks $150 in souping the family bus; that's not too bad, but the resulting performance increase seldom satisfies, and Louie can't sleep nights till he's sunk another $300 to get another 10 mph of speed.

Some can afford it — some can't. And we'd be the last person in the world to help deny Junior a new pair of shoes! So take your souping in easy steps to keep pace with your income. However, we will say that you can do a neat little job on the V8 for $125 which will give you a little more snap at all speeds. But don't expect your true top speed to go up more than 8%. Too many guys expect too much for too little. Every horsepower you add costs you money, and the higher you go, the more they cost.

Maybe we harp too much on costs, but we feel that they're an extremely important item in this souping business. It's one of the big reasons we're even doing this work, isn't it? Costs become even more vital for the simple reason that there are a million ways we can save — do big things with the same dollar that would do little things somewhere else. An experienced mechanic can sometimes get more "soup" with $100 than Leadfoot Louie can with twice that. So we're going to emphasize this subject of economy clear through this book, and try to direct you toward getting the very most for your souping dollar.

So now we're rolling. We've kind of gotten in the mood of the thing, we know what we're after, and we know more or less what we're up against financially. But the rewards are well worth it; the roaring of the wind at 100 mph — the throb of twin pipes at 5000 rpm — the back-breaking acceleration when you step down on it — pavement streaking under you at 170 ft. a second. Wow! It's terrific!

So come along and let's take a look at this fabulous engine we're going to work with.

A successful experiment in sprint car design, this rig has turned well over 120 mph in ¼ mile, from a standing start. A tubular frame and unique I.F.S. system are featured. Engine is an over-bored Mercury, uses nitro methane mixture as fuel. Note absence of radiator.

CHAPTER 2

THE V8 FAMILY TREE

IT ALL started back in 1932, the depth of a bad depression. John Q. Public was in a bread line, Marmon was saying what this country needed was another 16-cylinder car and Henry Ford's Model A was obsolete.

So what did the genius of Dearborn do? He ups and brings out a cunningly-simple 2,500-lb. 8-cylinder car and "scoops" the automotive world for the nth time! Millions of V8-powered cars have been sold since and today, that basic block leads the stock speed field, with figures like 300 hp, 6000 rpm, and 210 mph!

Why do we speed enthusiasts prefer this engine? We think the main reason originally was that, with the short crankshaft and light rods, the acceleration and speed of those early Ford V8's was phenomenal and the speed world sat up and took notice. The engine layout was perfect for souping. And then, parts were cheap and widely available. It wasn't long before special heads and manifolds were on the market, and they were pulling upwards of 200 hp before the war.

We know what happened since, and where it will all end .... well, let's move on — this is getting us nowhere. Before we get any farther along, let's briefly look over this line of Ford and Merc V8 engine models and see just what we have to work with. Here's a thumbnail sketch of each model from the beginning:

1932 — Model 18; As brought out, the "V8" featured its present layout of 90° "V," cast alloy iron block, side valves, gear-driven cam-shaft, solid tappets, etc. The bore and stroke were 3-1/16 x 3-3/4 in., giving a total piston displacement of 221.0 cu. in. Compression ratio was 5.5:1 and the peak output was a modest 65 hp at 3400 rpm.

The heads were the old 21-stud type of cast iron, 18-mm plugs, with the water pump units built into the front of each head. The cam-shaft gear was of fabric, bolted to the shaft. Valves had a head diameter of 1.54 in., 0.312-in. stems, and the spring pressure with valves open was 78 lbs; no seat inserts were used. Aluminum flat-head, 4-ring pistons were employed.

The rods were 7 in. between centers with insert-type floating bearings, running on 2.0-in. diameter x 1.94-in. crankpins. The crankshaft was a forged 90° 4-throw counterbalanced type with three 2.0 in. diameter poured main bearings. A cast iron 9-in. flywheel was used, weighing 39 lbs. The entire engine weighed around 525 lbs.

1933 — Model 40; Major change here over the above model was the adoption of aluminum cylinder heads of 6.3:1 compression, which raised peak power to 75 hp at 3800 rpm.

1934 — Model 40A; No major changes except the adoption of the now-

Fig. 2-1. External views of the current Ford V8 engine.

Fig. 2-2. External views of the current Mercury engine.

famous Stromberg "48" carburetor, a duplex down-draft type with 1.03-in. venturis and No. 48 main jets as standard (0.048 in. hole). Formerly the Ford carb had been a single-barrel type. The new duplex carb had an amazing effect on overall performance — raising the peak rating to 85 hp at 3800 rpm and substantially increasing low-speed torque.

1935 — Model 48; No major changes.

1936 — Model 68; Steel pistons were adopted on later models, with solid skirts and flat heads. Ratings remained the same.

1937 — Model 78; Several major changes were made on the Ford "85" engine in 1937, and this model can be considered the forerunner of the modern line of V8's. The aluminum heads were redesigned so that the water outlet pipe was in the center, with the water pumps moved into the front of each block bank. In conjunction with the new combustion chamber shape, the cast steel pistons were given domed heads to reduce knock by giving a better "quench" area; the changes also dropped compression ratio to 6.12:1.

Valve timing was cooled off just a bit to give better low-RPM pulling (power rating remained the same). Also in this connection, the Stromberg "97" carb was adopted, using smaller 0.97-in. venturis and No. 45 main jets. The same basic crankshaft was used, but diameter of the main bearings was upped to 2.4 in. and insert bearings were used instead of the former poured type.

1938 — Model 81A; No major changes except 14-mm plugs.

1939 — Ford 91A, Mercury 99A; With the expanding economy and rising standard of living which this country found itself blessed with in the late 1930's, Henry Ford decided that there was a lot of money floating around in the "medium" price bracket that he wasn't getting a crack at with his $600 Fords. And so the Mercury was born. The significance of this development as far as we're concerned here is that the entire V8 engine was redesigned so that the same basic power plant could be used in both cars.

The major difference in the engines used in the Ford and Mercury cars was that the Merc had 1/8 in. larger bore (3-3/16 in.), which gave it a displacement of 239 cu. in., compared with 221 for the Ford. Ratings of the new engines were 85 hp at 3800 rpm for the Ford, and 95 hp at 3600 rpm for the Merc.

Heads were again redesigned, this time in cast iron with 24 studs and water outlet at the center; compression ratio was 6.1 and 6.3:1 on the Ford and Merc respectively, due to the difference in their cylinder volumes. Camshaft and valves were unchanged, except the cam gear was pressed on. Block layout was the same, but an entirely new crankshaft was used; it was about 1½ in. longer and had 2.5 in. diameter mains, though they were a bit shorter. To ease rod and bearing replacement on the Ford, the crankpin diameter was held at 2.0 in., but the new Merc carried 2.14-in. pins on the same basic crank. Rods remained basically the same (7 in. length) and domed steel pistons were used in both engines.

1940 — Ford 01A, Mercury 09A; No major changes.

1941 — Ford 11A, Mercury 19A; No major changes.

1942 — Ford 21A, Mercury 29A; Heads were modified to raise compression ratio to 6.2:1 on the Ford and 6.4 on the Merc; this raised ratings to 90 hp at 3800 rpm, and 100 hp at 3800 respectively. Some ignition changes were also made, but the same basic layout continued.

Right here war intervened.

POST-WAR ENGINES

When peacetime production started again in 1946, the now-famous Model "59A" engine was brought out and used in identical form in both the Ford and Mercury cars. It was an improved version of the pre-war Merc, carrying the same bore and stroke (239 cu. in.) and rating of 100 hp at 3800 rpm. Compression ratio was raised to 6.8:1 to take advantage of higher-octane fuels, but bearing sizes, valve timing, and other major items were standard pre-war Merc. This 59A power unit was continued unchanged for both cars through 1948.

For 1949, the Ford Motor Co. made quite major modifications on the much-used 59A in an attempt to stay with the trend towards more HP and more flexible performance. The block was not greatly changed except that a stamped, sheet-metal bell housing replaced the cast-in section of the 59A. The heads were altered by moving the water outlets to the front, with a thermostatic by-pass unit built in at this point; this necessitated a small hole in the block here which must be plugged when using

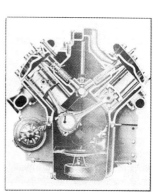 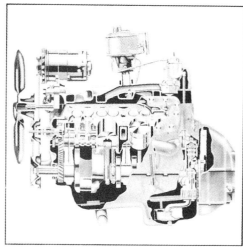

Fig. 2-3. Cutaway sections of the current Mercury engine.

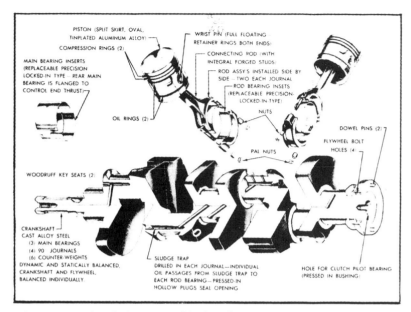

Fig. 2-4. Crank-rod-piston assembly for the late post-1949 Ford and Merc. Note the individual doweled rod insert bearings in place of the old floating sleeve.

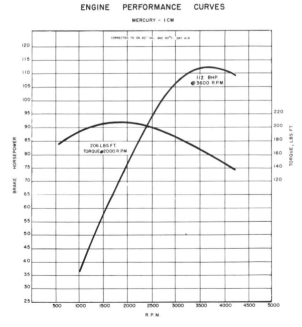

Fig. 2-6. Power and torque curves for the current Mercury engine.

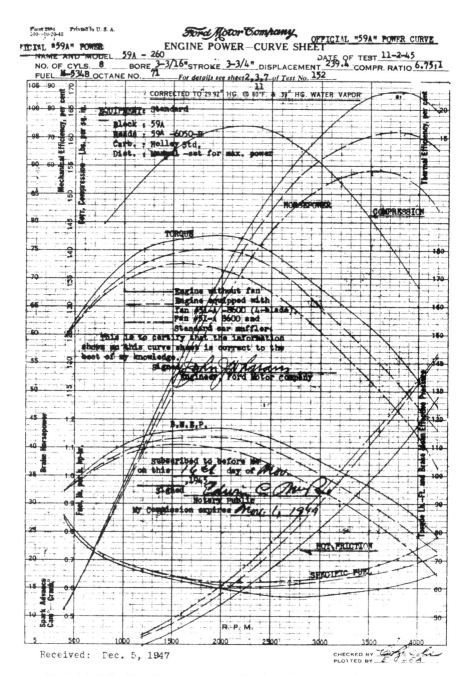

Fig. 2-5. *Official performance curve sheet for the 59A Ford-Merc engine, 1946-1948.*

earlier heads.

In the lower half, the crank was modified considerably, though bearing sizes remained the same. Counterweights were changed and individual doweled insert bearings were used for each rod end in place of the former single floating bearing on each crankpin (this also necessitated some changes in the oil passages in the crank). The Mercury crank was given a 4-in. stroke, which raised the displacement here to 255.4 cu. in.

Major changes were made in ignition. The distributor was placed at the front of the right head and driven through a shaft from a spiral bevel gear on the nose of the camshaft. The system was altered from the double-breaker type to a single breaker with 8-lobe cam, and utilized a full vacuum spark advance mechanism with no centrifugal component. This ignition was not good as far as the souper was concerned.

This basic engine, as brought out in 1949, was designated the "BA" series for the Ford, with compression of 6.8:1 and rated 100 hp at 3800 rpm. The "CM" series for the Mercury (255 cu. in.) was rated 110 hp at 3600 rpm. These engines have been continued without major changes up to the present.

So there are all your Ford-Merc V8 engines, from the little 65-hp Model 18 of 1932 to the big 112-hp 1951 Merc. And it's a list that leaves the true souper's tongue out like a necktie!

Getting down to practical cases, just what are the best models for our purposes? You may not have much of a choice if you happen to own a '34 coupe and have only peanuts to sink in your souping, but here is how they stack up generally from a strict souping standpoint:

With the present trend toward large bores and strokes, pre-war Ford blocks (221 cu. in.) are no longer very popular. As a matter of fact, there isn't much available in heads for the pre-1939 21-stud layout, and these blocks are not the first choice anyway because of smaller bearing sizes. So we find the pre-war Mercs and all post-war engines at the head of the souper's list. For all-around purposes, the famous 59A block has been found best of all.

So here's the raw material we have to work with. It'll do wonders at the clutch if we handle things right. Now where do we go from here? Before we go anywhere, let's take a quick look at the basic fundamentals of engine performance so we can know which direction to take when we do get started.

So come along into the land of dynamometers, slide rules, flow meters — and burned-out engines!

Chassis view of the Floyd Clymer Motorbook Special, during *initial construction*, in the shop of Kenz & Leslie, Denver. Note straight shaft connection between the two Merc engines, and quick-change gearbox. Frame is tubular, suspension is by torsion bars. Many months of work and many dollars went into the building of the record holder... (America's Fastest Car and the World's Fastest Hot Rod.)

CHAPTER 3

PERFORMANCE FUNDAMENTALS

WE ASKED Leadfoot Louie the other day why he always used such low gear ratios and wound his engines so tight in all his cars. His answer nearly floored us: "Why, you crazy ike — don't you know your horsepower goes up as you increase your RPM? The faster you turn 'er, the more power you get — as long as she'll stay together. I gear my Merc to wind 6000 right along. Can't understand why I never win anything on the track, though — must be my chassis!"

Oh yes, Louie's an engineer all right — and he's got the broken crankshafts and wrecked blocks to prove it. But let's us avoid Louie's fate and look a little more closely at this important item of performance before we try to establish our basic souping steps. If we expect to be able to intelligently plan our souping procedure, then we've got to be able to put a "price tag" on each step. And to do this, we've got to understand the fundamentals of engine performance and just what's going on in that engine when it's "performing."

In the first place, what does performance include? The big item here as far as we're concerned is power. But when we think of HP, we usually think only of the peak output — like "my souped Merc pulls 180 hp at 4500 rpm." But with a three-speed transmission in street driving, what that engine develops at 1000 rpm is mighty important too. In other words, performance includes an engine's HP at all RPM.

And then, it includes things like fuel consumption, friction, efficiency, etc. These may not seem immediately vital to our souping work, but we'll find they all play a part. So let's swing right into the business:

POWER AND TORQUE

Every amateur mechanic speaks glibly of horsepower, but not many know what power really means. Here's the simple definition: POWER IS THE TIME RATE OF DOING WORK.

You can lift a 60-lb. crankshaft up 3 ft. to your waist, and you've done 60 x 3 = 180 FOOT-POUNDS of work. But this doesn't tell a thing about the actual power you've produced. Your power output, then, depends on how FAST you can lift that crankshaft from the floor to your waist.

The U.S. horsepower is officially defined as 550 ft-lbs. of work PER SECOND. In other words, if you can lift the crankshaft up 3 ft. in one second, you're producing 180/550 — about 1/3rd of one HP. (We don't know how Charles Atlas was rated, but we doubt if he could've raised the crank in less than ½ second!) Or similarly, suppose your car is rolling 90 mph (132 ft./second) and the total drag or resistance is say 300 lbs. That means your engine is pulling 300 lbs. through 132 ft. in one second, which is 300 x 132 = 39,600 ft-lbs. of work—and it is producing 39,600/550 = 72 hp. Simple?

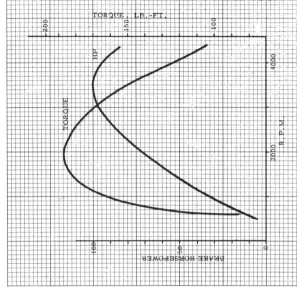

Fig. 3-2. *Typical dynamometer test curves for a stock car engine.*

Fig. 3-1. *Typical engine performance test setup, using a "water brake" dynamometer; note the scale on the right to measure the torque reaction of the water turbine unit in its cradle.*

Now we come to "torque" — which is really a simple thing, but it sometimes leaves the amateur mechanic in the dark as to its true relationship to HP. Torque is a FORCE — not work. It is the word we use to denote the twisting or turning effort exerted about a center of rotation. In other words, if you take a crank with an arm 1½ ft. long and turn it on the end with a pressure of say 30 lbs., then the amount of torque you're exerting is 1½ x 30 = 45 POUND-FEET.

But this isn't work because you'd still be exerting the force whether the crank was turning or stalled. (An interesting example of this is a gas turbine engine; it develops its maximum torque with the power turbine stalled — but the HP output is, of course, zero!) But now if you actually turn the 1½-ft. crank one revolution, then your hand has traveled 7.87 ft. and has pulled 30 lbs. through this distance — which is, of course, 7.87 x 30 = 236 ft-lbs. of work. (On the other hand, if we figure with the torque value of 45 lb-ft., our distance for one revolution with a 1-ft. radius would be only 6.28 ft., but the final answer would be the same.)

In other words, torque must turn through a certain distance, say one revolution, before it becomes work — and how fast it turns, of course, determines how much power is being produced. Torque is related to the U.S. horsepower by the following formula:

$$HP = \frac{T \times RPM}{5250}$$

Take your slide rule and check any of the HP and torque curve sets in this book and you'll see that they're exactly related by this formula.

PERFORMANCE TESTING

We've dwelt at some length on these power and torque definitions because it's essential to our work that we thoroughly understand the power and torque "curves." Most of you know the procedure in taking test data for an engine on a dynamometer. Here it is briefly:

A dynamometer is a machine for artificially loading an engine and measuring its torque output. The "load," or resistance to turning, is adjustable and can be produced with electricity, churning water in a turbine, or a brake band on a drum; with the resistance section absorbing the output of the engine, the torque reaction of the dynamometer unit in its cradle is measured to calculate the torque being produced.

In running a power test, the load of the dynamometer is carefully adjusted so that it will just hold the engine speed to say 500 rpm with the throttle wide open; the torque is then read and the HP calculated from the above formula. This procedure is repeated for speeds of 1000 rpm, 1500, 2000 and so on till the full curve is drawn on the graph. Fig. 3-1 shows a typical engine test layout using a "water brake" (a dynamometer is sometimes referred to as a "brake," because actually that's all it is). Fig. 3-2 shows the test curves you might get from a typical stock car engine.

The first thing you'll probably ask is, "How in the world can the HP

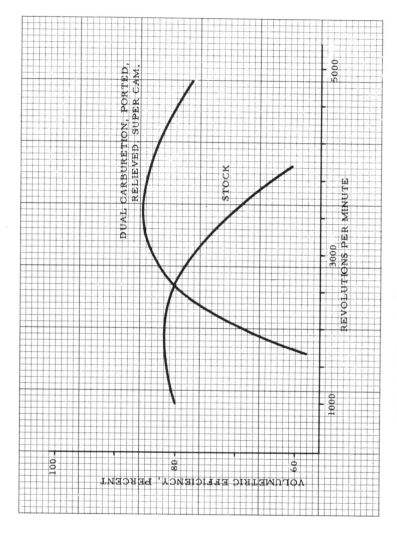

Fig. 3-3. Volumetric efficiency curves for the Ford V8 engine.

be going up while the torque is dropping off?" Good question. It's all a matter of RATES. Since HP is directly proportional to both torque and RPM, it's obvious that, if the relative drop in torque between two RPM points is not as great as the relative increase in RPM, then the power will be going up; when the two rates are equal, the HP is at the peak. For example, from Fig. 3-2, between 2500 and 3000 rpm, the torque drops from 181 to 169 lb-ft., or $6\frac{1}{2}\%$ — but the RPM increase is $3000/2500 = 20\%$. So the power is upward.

THE TORQUE CURVE

Thus we find that the torque curve reaches a peak at about 1/2 the peak power RPM and drops off on both sides. Okay — but why? First consider the drop-off at higher speeds. The reasons here should be quite obvious: (1) Decreased "breathing" through the valves, and (2) increased friction in the engine.

We all know the effects of RPM on breathing, or "volumetric efficiency," through the carb, manifolds, valves, etc. Volumetric efficiency refers to the volume, or cu. in., of fuel-air mixture at atmospheric pressure and temperature which the cylinder actually draws in at a given RPM (with full throttle), as a percentage of its displacement. Fig. 3-3 shows typical volumetric efficiency curves for the Ford V8 engine in stock form, and then converted to dual carburetion, ported, relieved, and with a modified cam. Quite a difference!

The main reasons why volumetric efficiency drops off at high RPM are because of gas turbulence in the manifold and ports, skin friction of the gas against the channel walls, and pressure losses caused by contracting the flow column in the carburetor, ports, and valves. These effects are killing — and increase about as the square of the flow rate! As a result, we find the effective cylinder pressures on combustion (which determine the torque) dropping off in a hurry above 2500 rpm.

And then there's engine friction. This is a killer too because it also increases approximately as the square of RPM — and there's not a thing we can do about it. If our engine loses say 10 hp in the bearings, pistons, gears, accessories, etc. at 1500 rpm, it will be losing around 90 hp at 4500 rpm! So you put together the effects of friction and breathing losses, and you can easily see why our torque drops off so sharply at high RPM.

Now how about that drop at low RPM? Certainly we can't blame that on friction and breathing. It's a bit harder to explain here, but the drop is due to two main effects: (1) Valve timing, and (2) decreased compression pressure.

In the matter of valve timing, we face a compromise, whether we like it or not. If we're to have fairly decent breathing on the intake stroke in the high speed range over 3000 rpm, we've got to have that intake valve open for 50 or 60° of the next up-stroke of the piston, to take advantage of the inertia or momentum of the incoming flow of gas to further fill the cylinder. This deal is wonderful at 4000 rpm, but at 1500 rpm, the cylinder could get a full gulp anyway, and the piston just pumps mixture back into the intake manifold on the up-stroke — and compression can't begin till the valve is closed (in fact, on a 7:1 compression ratio engine,

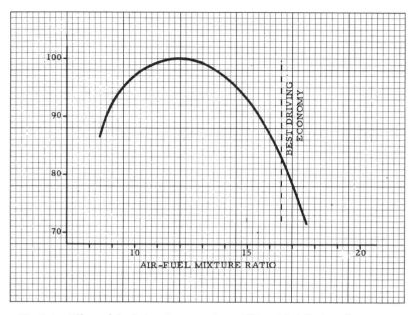

Fig. 3-4. *Effect of fuel-air mixture ratio on HP, with full throttle, constant RPM and spark setting.*

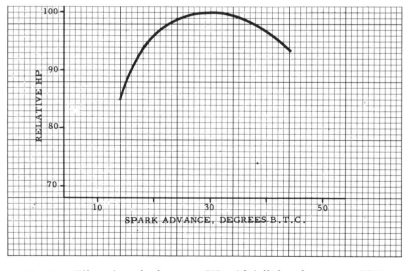

Fig. 3-5. *Effect of spark advance on HP, with full throttle, constant RPM, mixture, ratio, etc.*

this late valve closing has the effect of reducing the effective compression ratio to 6.2:1). But we have no choice rather than face this compromise, so we must be very careful in choosing our valve timing to fit our performance requirements. We'll discuss this more later on.

Decreased compression pressure is another bug at low RPM. The late valve closing mentioned above is largely to blame, but there's also a loss of compression heat. At high RPM, the compression is done so fast that there is no time for the heat generated in the mixture by this compression to leak away through the cylinder walls; therefore it is retained in the mixture and gives a much higher final compression pressure. At low speed, however, some of this compression heat has time to leak away and our final pressure suffers.

B.M.E.P.

There is still another concept of engine power that might help us to better picture souping procedure. You'll recall in our earlier HP formula that the torque was merely a force — but if it moved through a distance, one revolution, it became work — and we thus combined it with RPM to obtain our HP.

Now if we consider the average gas pressure pushing down on the piston during the power stroke as our force, and the length of the stroke as the distance through which this force acts, then we can let this value replace torque in the HP formula. But to make it simpler, let's just use the pressure on one sq. in. of the piston head, then combine the total piston area with the stroke, which will give us the piston displacement, and we now come up with another simple formula for HP:

$$HP = \frac{P \times D \times RPM}{792,000}$$

where "P" is the average cylinder pressure during the power stroke which shows up as power at the flywheel (we call this "brake mean effective pressure"), and "D" is the total piston displacement in cu. in.

This enlightening little formula shows us that HP increases directly with these three factors. Double any one of them, and you double the HP — or raise each by only 26% and you double HP!

Let's take an example: On the stock Ford V8, our displacement is 239 cu. in., peak power is 100 hp at 3800 rpm, and peak BMEP works out to 87 lbs./sq. in. Now if by careful souping, we can raise peak RPM to 4500, BMEP to 125 lbs., and displacement to 267 cu. in. (1/8 in. bore and stroke), our HP jumps to: 125 x 267 x 4500/792,000 = 190.

This is all by way of illustrating the importance of these factors of cylinder pressure and piston displacement that didn't enter our original torque formula. We wouldn't have brought this new formula into it — since it doesn't simplify things — but we want to get it across to you how vitally these factors effect HP.

TWO FLIES IN THE OINTMENT

All our previous discussion has assumed that our hypothetical engine has been running under the best operating conditions from a HP standpoint. There are a thousand little things that can upset our "best conditions," but two very important items are fuel-air mixture ratio and spark advance. When running a dynamometer power test, the carburetor metering characteristics are generally left as stock, and the spark advance is handled by the automatic advance mechanism in the distributor. That's all well and good, but hot rodders — especially our Leadfoot Louies — have a fierce tendency to modify principles that our engineers have spent 50 years developing! So lest you get too frisky with mixture ratio and spark advance, let's see how they effect HP:

The theoretically correct mixture ratio for burning gasoline in a closed cylinder is about 15:1. But because the evaporation of the fuel cools and contracts the mixture in the manifold, so that the piston draws more of it in on the intake stroke, we find that our maximum torque, other factors equal, is developed with a mixture some 20% rich, or about 12:1. Fig. 3-4 shows this effect.

And it also shows that power and fuel economy don't exactly go hand-in-hand. Cars have been run with 21:1 mixture ratio (with special equipment), but cooling becomes a problem on such lean mixtures. With your engine you'll probably be more interested in performance than gas mileage anyway, but don't get the idea that you can go on adding bigger and bigger jets in your carbs and get more and more HP. We'll deal with this more thoroughly later on.

Your engine power is also very sensitive to spark advance. Fig. 3-5 shows this effect. The reason is obvious: The mixture does not burn instantly, but requires a definite time, depending mostly on cylinder turbulence and pressure. But also, for maximum efficiency, we must see that the combustion is nearly completed by the time the piston starts down on the power stroke. This requires that we be able to vary our spark advance from about 5° to 30° before top center to get the maximum power under different combinations of load and speed. Stock spark advance mechanisms do a fairly good job, but we may expect to have to do some modifying here. We won't go into this any further right here, but we felt the general subject warranted emphasis in this chapter on performance.

FUEL CONSUMPTION

This important item in the overall performance picture is probably conspicuous for its absence in this chapter. We dealt with fuel consumption at length in our book, "Souping the Stock Engine," to keep the story complete, but we feel right now that the subject is of absolutely no importance to the souper, so we're not going into it. However, if you're the one in a hundred who's interested in strict economy as well as performance, we'll say this to steer you in the right direction:

For the maximum possible fuel economy, use a good manifold "hot spot," go to smaller carb jets, and run your engine at the lowest possible RPM under all conditions. Your minimum fuel consumption, on a basis of lbs. per HP per hour, comes at medium RPM and at full throttle — but on the road, you'll find better gas mileage if you can lug your engine clear down to 1000 rpm. This, of course, means overdrive and a high axle ratio. It's a rough combination from the standpoint of acceleration, but it's what you'll have to pay for economy.

So that about concludes this brief review of general engine performance. Perhaps we spend too much time on basic theory, but if you expect to attack your souping with some system and intelligence, you've got to know just what you're after. We've shown in this chapter that our aims are not really very complicated — actually we're after three things: (1) Maximum cylinder pressure or BMEP at all speeds, (2) maximum displacement, and (3) maximum possible peak RPM (not just RPM, as Louie thinks!).

Our next job is to establish the basic steps we can take in these directions — and most important — pin a price tag on them. So come along and we'll rip that stock V8 apart and fix it up so that Aunt Effie won't get within twenty feet of it!

An interesting, though unsuccessful adaptation of double overhead camshafts (4) on a Mercury block, as installed in the Bob Estes Spl., entered at Indianapolis, 1951. Note gear-driven shafts, fuel injection system.

Fuel injection has already made its appearance in the field of hotrodding, and has even stood the test of the Indianapolis "500" Race, having been used on a number of cars both during qualifying trials and during the race itself. The car pictured is an early attempt at fuel injection, built in 1947 by Stuart Hilborn, of Los Angeles, and utilizing a 1934 Ford V8 engine. The Hilborn-Travers Engineering Company are now producers of special fuel injection equipment for racing and competition use.

CHAPTER 4

PLANNING THE JOB

OUR LEADFOOT Louie has already blown that block he had going in Chapter 2! He went full-race with it, but naturally didn't bother to boost oil pressure or have the crank set balanced after he bored and stroked. First time out, he tried to get 'er up to 7000 in low gear — and that was it. They tell us all he salvaged was the rear axle!

Louie, of course, has two strikes on him from the start but we've also seen some blunders at the hands of the intelligentsia. There was the brilliant character who ran aviation triptane in his track job and burned up a perfectly good set of pistons. Or there's the boy who sank $450 in super-race equipment for a stock-displacement street roadster — he could barely keep the engine running in traffic and today, he's got the whole kaboodle for sale.

Lack of planning hexed these guys, and we'd like to do something about it right now so such things won't be happening to any of our readers.

THE PATHS TO POWER

In the first place, before we can plan our souping procedure, we must know exactly in what general directions we're heading and what specific souping steps will give us the most for our souping dollar. In the last chapter, we analyzed performance and firmly established the basic factors that determined our HP. Since HP is all we're really after, our next job is to break down the three factors — that is, effective cylinder pressure, piston displacement, and peak RPM — into basic souping steps.

First, take a look at our HP formula again:

$$HP = \frac{P \times D \times RPM}{792,000}$$

The displacement and RPM factors are specific values, but obviously that "P," for effective cylinder pressure, depends on two distinct sub-factors: (1) The weight of fuel-air mixture drawn in on the intake stroke, and (2) the efficiency with which it's burned.

Right here, we come to a vital element in all our souping work — and that is GAS DENSITY. Specifically, density means weight per unit of volume, or say lbs. per cu. in. It depends on temperature as well as pressure. The density formula for air is:

$$D = \frac{460 + T}{2.7 \times P} \quad \text{(lbs./cu.ft.)}$$

where "P" is the atmospheric pressure in lbs./sq.in., and "T" is the temp in °F.

It doesn't mean a thing to say that an engine consumes 250 cu. ft. of air per minute. In the thin air on Pikes Peak on a warm day, your total fuel-air mixture at this rate would weigh about 12 lbs. In freezing weather at sea-level, it would weigh 21.9 lbs. And if you were burning alcohol, its evaporation would drop manifold temperatures to near zero, and you'd consume even more lbs. — though the same number of cu. ft.

Experiments have shown that HP is directly proportional to the weight of air burned per minute, assuming constant RPM, fuel-air ratio, spark advance, etc. So we see that it's the WEIGHT of mixture — not just cu. in. — which we draw in on the intake stroke that determines our HP.

Keeping this in mind, then, let's set down our five basic paths to power: (1) Increase piston displacement, (2) Increase the weight of fuel-air mixture drawn in on the intake stroke, (3) increase the efficiency of combustion, (4) decrease engine friction, and (5) increase the peak RPM.

Okay — that's simple enough. Now let's break these basic steps down further into individual souping steps on the Ford-Merc V8 (incidentally, we'll combine the last two steps into one category, since what helps one will automatically help the other):

FOR INCREASING PISTON DISPLACEMENT
1. Boring and stroking

FOR INCREASING WEIGHT OF MIXTURE DRAWN IN
1. Pull air as cool as possible into carbs and use no manifold "hot spot"
2. Increase total carb throat area
3. Increase flow path area in manifolds and ports
4. Smooth surface of the gas flow path
5. Have as straight a flow path as possible
6. Increase valve discharge area
7. Modify valve timing and rates
8. Enlarge and smooth gas flow path from valve into cylinder
9. Enrich fuel-air mixture, within limits
10. Burn fuels with very high latent heat, such as alcohol
11. Supercharge

FOR INCREASING EFFICIENCY OF COMBUSTION
1. Raise compression ratio
2. Use aluminum cylinder head
3. Proper spark advance and sufficient spark voltage
4. Proper fuel-air mixture ratio for maximum power
5. Use fuel which will allow no detonation or pre-ignition
6. Have good exhaust scavenging (open pipes, etc.)

FOR DECREASING FRICTION & INCREASING PEAK R.P.M.
1. Use lighter pistons and rods
2. Balance the crank assembly
3. Use fewer piston rings
4. Increase bearing and piston clearances
5. Increase oil pressure
6. Use heavier oil (sometimes)

LIMITATIONS

Now in attacking our problem, there are three definite limitations

Fig. 4-1. A typical Navarro semi-race road engine. A deal like this will run somewhere between $350 and $500, not including labor, and will generally pull 160-180 hp on pump gas.

which we must consider:

FUEL — Though a competition engine is not generally limited in this respect, a road engine must be able to operate satisfactorily on fuel available at any filling station. Premium (Ethyl) gas rates now at about 80-82 octane, and this will limit compression ratios to about $8\frac{1}{2}$:1 if we expect to do much full-throttle running (more about this later).

MUFFLERS — Most states have pretty strict laws on muffling, so we'll have to figure on some silencing arrangement for our road engine. Stock exhaust systems are awfully rough on HP, but dual straight-through systems are widely available for the V8, and these are very efficient and practical.

LOW-R.P.M. PERFORMANCE — To make a long story short, if we're going to do much city driving with our rod, we're going to have to be mighty careful about cam timing. Safety requires that we have some 15-20 hp available at 1000 rpm, yet some of the "grinds" will barely idle at this speed on a stock-displacement engine! Many "rodders" have been roughed up because they thought they could get across a crowded street — only to have their super-race engine backfire and kill when they clutched it.

We'll discuss this more later, but we want to list right here that cam timing is a definite limitation on a road engine.

FOCUSING ON COSTS

That takes care of the "bugs." We're ready now to get down to some details. First let's set up some souping categories. In line with our policy of stressing economy in our souping, we're going to do this on a total cost basis. Assuming you already have your stock V8 to work with (in pretty good shape) and that you can do some of the work yourself like assembly and fitting — we'd set up the following four engine categories with the Ford-Merc V8 block: (1) Conservative road engine, $125; (2) medium road engine, $250; (3) hot road or track engine, $500; and (4) super competition engine, $1,000.

Perhaps you'll disagree with these cost figures. Maybe you think we should allow $200 for the conservative job and $350 for the medium. But quick as you say that, some Leadfoot Louie steps up and asks, "How can I soup up my engine for $15?" So we think these cost limits will set a fairly good pattern for the average rodder — and, of course, you can always juggle the exact procedure to fit your pocketbook.

For you fellows who aren't too familiar with this hot-rodding business, here's a brief list showing roughly approximate prices for the major souping equipment and machine work on the V8 block (not including postage, taxes, etc.): Note—Prices may fluctuate from time to time.

Top-grade aluminum high-comp. heads	$ 75
Second-grade aluminum heads	60
Iron high-comp. heads	30
Overhead-valve head sets	500-750
Top-grade compound manifolds	45
Second-grade compound manifolds	35
Carburetors	12
Reground camshafts (exchange)	30
Adjustable tappets	14
Porting and relieving	35
Boring	20
Top-grade special pistons	45-55
Second-grade special pistons	20-40
Stroking a crankshaft	25
Stroking with "metal spray"	50
Chopped stock flywheel	15
Special aluminum flywheel	25
Complete balancing job	35
Exhaust headers	30
Special stock converted ignition	20
Special dual-coil ignitions	45-65
Magnetos	100-130

Now to some specific procedure:

CONSERVATIVE ROAD ENGINE — $125

Here's what we would suggest here:

Fig. 4-2. Tattersfield super-race setup, bored and stroked, quad manifold, and a "full house". With 295 cu. in. displacement, this engine develops 228 hp at 4200 rpm, on alcohol.

1. Aluminum 8½:1 heads (second grade)
2. Dual manifold (second grade)
3. 3/4 reground cam

This is the category for John Q. Public who just wants to soup up the family bus a little. The cam grind is mild to give good low-RPM lugging, and the whole setup is mild enough so that it should not require special ignition — if Mr. Public doesn't want the "long wind" to 6000 rpm! We've suggested second-grade equipment here. Why not? You're not competing with the champion! Buying cheap stuff does involve risks, however, as we'll point out later, so be careful. If you just don't want to use the second-grade equipment, omit the manifold. The peak output of the 239-cu. in. block with the above setup would be around 140 hp at 4200 rpm.

MEDIUM ROAD ENGINE — $250

We would suggest the following changes:
1. Aluminum 8½:1 heads (top grade)
2. Dual manifold (top grade)
3. Mild "race" grind cam
4. Ported and relieved
5. Dual-coil ignition
6. Chopped stock flywheel

This souping job is for the average rodder who has a little more money

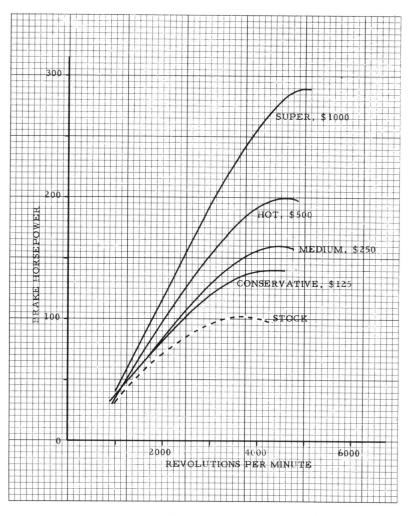

Fig. 4-3. What our souping dollar can bring us in the way of HP in the four categories; note that low-RPM output hasn't been helped much—you can't have everything!

to put into the job than John Q. P., who wants the most for his money, but still can't go too deep. You should be able to get away with a mild "race" grind cam here, but you'll need to handle the gear-box and clutch with some ginger! We definitely recommend chopping your stock flywheel for this job, rather than fitting an aluminum racing type, as the latter doesn't stand the pounding of the clutch plates too well over long periods of time.

Dual ignition will be necessary here too, as we are getting into cylinder pressures where the stock ignition setup will likely cut out on full throttle in the 4000-5000 rpm range. Your peak here should be about 160 hp at 4500 rpm.

HOT ROAD OR TRACK ENGINE — $500

You can buy a lot of "soup" for $500 if you work it right. Here's how:
1. Aluminum $7\frac{1}{2}:1$ rated heads ($8.8:1$ final ratio); use higher compression with special fuel for competition
2. Triple manifold
3. "Full-race" cam grind
4. Ported and relieved
5. 3/16-in. bore and late Merc crank (4-in. stroke)
6. Special pistons, type depending on use
7. Increased piston and bearing clearances
8. Special hard rod bearings
9. 60 lbs. oil pressure
10. Dual ignition
11. Chopped flywheel (aluminum for track)

This is about as "hot" as you'll want to go for a road engine and, in fact, it's got plenty of stuff for track work (though as in all competition, the best is none too good). We've jumped to nearly 9:1 compression ratio as the absolute top with premium pump gas. Also, with our large displacement here (286 cu. in.), we can get away with (and in fact, we need) three carbs and a full-race cam. This wouldn't do with stock displacement, though. Our pistons, bearings, and clearances will depend on whether we're building up a road or track engine, and we'll discuss this later.

Special ignition is absolutely essential here, but we needn't necessarily go to the magneto. The chopped flywheel is desirable for better acceleration — the aluminum wheel is suggested for racing. All in all, this engine should give fairly good low-RPM performance for street driving (though it won't outpull your stock setup below 1500 rpm), and the peak in road trim will be around 200 hp at 4600 rpm.

SUPER COMPETITION ENGINE — $1,000

You won't be needing to worry about how she's going to get around in traffic with this one, so you can shoot the works. And you can do wonders with $1,000! Here's how we would spend it:
1. Overhead-valve heads, 14:1 ratio
2. Triple or quad manifold
3. Super-race or "mushroom" track grind cam
4. Bore and stroke according to displacement regulations, but get 296 cu. in. if no limits

5. Three-ring solid-skirt racing pistons
6. Increased clearances
7. Special bearings
8. 60-80 lbs. oil pressure
9. Magneto ignition
10. Aluminum 12-lb. flywheel

This isn't all you can do on this one by any means. It's a good start — but don't be afraid to throw the book! It's only by doing the impossible that these backyard rodders have brought us to where we are today. Incidentally, if you've done half a job on this "super-competition" engine, you should be able to pull a good 0.95 hp per cu. in. at 5000 rpm on alcohol.

COST VS. RESULTS

There are your four souping categories and the specific souping steps in each.

Now you may be wondering why we chose these particular steps as we did. In a word, it was our attempt to get you the most for your souping dollar. Don't get us wrong — every souping step from adding high-compression heads to polishing your rods is important when you're after that last HP. But some steps are a good deal more important than others when you've got a limited amount to spend and you want to get the most soup for every dollar you sink.

Let us illustrate: Suppose you have a late Ford V8 and you want to do a little souping. Your friend, Joe Blow over in Podunk, is running a big-bore stroker mill in his car, and he claims that the increased displacement gives him "terrific stuff" at all RPM. That testimony might not stand up in court, but it's good enough for you, so you sink $130 in a 1/8 x 1/8 in. bore and stroke job. Now what have you got? Your displacement increase is 12%, and your peak HP boost would be about the same — so you've spent about $130/12 = $10.80 for each percent increase in HP.

You could have done better. A 3/4-reground cam with adjustable tappets would have cost you only some $50, and your HP increase would have been around 15% - or $3.33 percent. Get it? Of course, increased displacement will give you much better low-RPM performance than a reground cam, and if that's what you are after, don't go for the cam. We'll go into all these details later, but we want you to get the general idea of this souping economy right now. It can save you money in a thousand ways.

For you fellows who are primarily interested in costs (and there must be a lot of you), perhaps it would be a good idea right here to list the major souping steps IN THE ORDER OF THEIR TOTAL COST FOR EACH PERCENT INCREASE IN PEAK H.P. The most economical step is at the top, graduating down to more costly methods — and remember, this list applies only to the Ford-Merc V8 block:

1. Reground cam
2. Compound manifolding
3. High-comp. aluminum heads

4. Boring 3/16 in.
6. Late Merc 4-in. crank
7. Stroking stock crank

This listing is pretty definite on a strict basis of cost and peak HP, and leaves little room for argument. However, you may not be interested primarily in peak HP, and that would hex the list somewhat. For example, if you just had to have increased output clear through the whole RPM range, a reground cam would be of no use to you — nor would dual carbs, porting, or relieving. Your order then would be: (1) Boring, (2) high-compression heads, and (3) late Merc crank. So don't blow your stack if this list doesn't suit your case; we'll get to you later on.

PLAN AHEAD

We'd like to mention here also the economy of planning ahead. Don't get all hot and bothered and buy a lot of equipment for a medium road engine today, then tomorrow decide to go "super-competition." A dual manifold is nice for the road, but you can do a lot better for the track. A 3/4 cam is fine for street work, but has little effect feeding big bores and strokes. And what good are 8½:1 flat heads if you really want overheads?

In other words, plan as far ahead as you possibly can. Suppose you have $150 to spend today, but eventually you expect, and are more than willing, to sink $500 in a hot road engine. In this case, you'd best modify all plans. It might be best here to bore out first, fit the oversize pistons, then add a full-race cam; add special ignition next, then go for heads, manifolds, etc. Get the idea? More on this later.

WHO DOES THE WORK?

Don't look now, but it's going to cost you two or three bucks an hour to have your souping work done for you! And there are hundreds of shops all over the country that are just rarin' to do anything on your engine you want them to. Their prices are reasonable and they'll do a good job—but this idea isn't saving you money. So we would like to suggest that you do as much of your own souping as you feel capable of doing.

There are two good reason: You'll save money — AND you'll gain a lot of invaluable experience through an intimate knowledge of your engine. Mechanically speaking, you'll "grow up with your block." The guy who buys his engine in a box is the first one to drive the daylights out of it! When you do some of your own work, you're going to know how your engine is set up, what it will take and what's most important — how to improve it.

We realize there are a lot of things you likely won't be equipped to do. You can't grind a crankshaft, but that doesn't mean you have to throw down your block in the machine shop and say, "Pull the crank and stroke it 1/8th." YOU PULL IT! That's about like Leadfoot Louie a couple of years ago when he first started souping. He drove his stock Ford sedan up in front of the Joe Doakes Speed Shop, dashed into the office, and asked for an eighth-inch stroke! Anyway, we want to emphasize this idea of doing your own work — for your own good as well as your pocketbook's.

Now just how much can you do? We well realize that 99 out of 100 "speed tuners" don't have access to tools and shop equipment beyond bare essentials (wrenches, hammers, screwdrivers, etc.). And since additional tools and equipment could make your engine investment look like chicken feed, we'd be the last to suggest that you fit out a machine shop. Souping machine work is a job for an expert anyway. So we'd only recommend that you invest in a good set of hand tools. They're absolutely essential in this work, and the investment will pay dividends for a lifetime.

We'd also put forth the suggestion of an electric drill, with a rack for setting it up as a drill press, and a flexible-shaft grinding attachment. This outfit will be invaluable in your chassis and body work, and will allow you to do your own porting and relieving and numerous other small jobs. It's worth considering, though by no means essential.

We could go on here with a lot of guff about keeping your tools neat and orderly, in good condition, etc., but you should know all this. We would, however, suggest that you keep careful RECORDS. Get a notebook and write down those measurements, tolerances, weights, etc., that you come across in your engine rebuilding. Keep close track of your spark plug and carb jet experiments. These records will almost surely prove invaluable to you one day.

So that concludes the first section of this book—the basic fundamentals of this V-8 souping problem. If you have waded through four chapters with us, you are probably beginning to wonder when we get to the meat. So come along and we'll work on the block.

CHAPTER 5

FITTING UP THE BLOCK

THE BLOCK setup is the most vital factor in the success of your souping efforts!

Too many guys give this subject only minor consideration in their "horsepower pipedreams." Anybody can stick in a reground cam, whack fifty-thousandths off the heads, or add a carb. But it takes a Leadfoot Louie to bore out 5/16ths and crack the cylinder walls — or neglect to rebalance after a bore and stroke job, and see his engine come apart at 6500 rpm — or run stock bearings and 30 lbs. oil pressure with a super-race track setup and have all kinds of lower-end trouble — or just stroke when he should bore, and waste $100!

So let's avoid Louie's blunders and give a little more attention to our block details. Our efforts will pay off in endurance and reliability as well as HP. (Incidentally, we are not going to deal with porting and relieving in this chapter on block work, as this subject properly belongs in a later chapter on the induction system.) So now that we have taken care of the preliminaries, let's dive right into this vital subject:

SELECTING A BLOCK

If you're one of the soupers in the "conservative" and "medium" categories — in other words, if you are just souping up the bread-and-butter vehicle that you drive to work, you won't have much choice. Your block is already there — soup it or leave it!

You should, however, put it in good condition before you soup up. The cylinder bores should be checked and, if they're worn over 0.002 in., they should be bored to the next standard oversize and fitted with new stock pistons and rings. All bearings should be replaced if worn, and the crankshaft should be checked for "out-of-round" and taper, and ground undersize if any journal is over 0.001 in. out. Also check the rods for alignment. If you've taken good care of your engine, driven fairly easy, and don't have over say 25,000 miles on it, you shouldn't have to worry about these items. (In fact, unless you're in serious doubt about the condition of your engine, it would hardly be practical to tear it down just to inspect it for a very mild souping job.)

For the hot road and competition conversions, you'll probably be building your car from scratch, and you might very well start with a brand new block too. At least the rods should be new. We don't like to insist on a big investment in a lot of new stock parts; you pay plenty for special stuff as it is. But just remember that machinery gets "tired" — no matter how much you've reground or refaced or reset. Fatigue cracks set in the metal and, before you know it, a block or a rod or a crank has fractured. In other words, don't take a block that's been on the road ten years and has 150,000 miles on it, and try to pull 220 hp from it. You might — and you might not. Why take the chance? If you have a fairly new block, okay.

INCREASING THE BORE

Increasing piston displacement is a basic step toward more HP, as we learned in Chap. 4, because this increases the weight of fuel-air mixture drawn in on the intake stroke. But the most important effect of increased displacement as far as our souping work is concerned is this: IT IS ONE OF THE FEW STEPS WE CAN TAKE THAT WILL BOOST THE H.P. AT LOW ENGINE SPEEDS! On a road car, our low-RPM output is vital, since we must have some acceleration at very low speeds for street driving.

An effect of increased displacement is to boost HP (or torque) over the whole speed range — but especially at low speed. Fig. 5-1 shows what we mean; it shows differences in the power curves resulting when peak HP

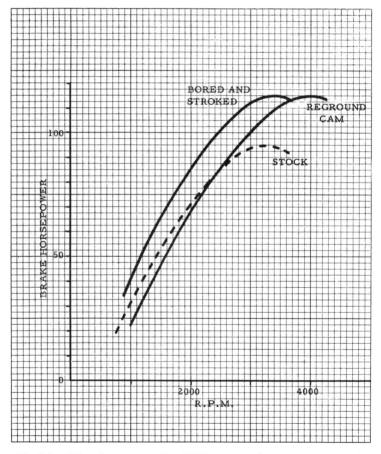

Fig. 5-1. What happens to low-RPM output when you raise HP by 15% with a reground cam, or by increasing piston displacement.

is boosted 15% by increased displacement and by a reground cam. See what we mean by low-RPM output? So remember, increased displacement is vital if you're after an engine with lots of guts at low speed.

At any rate, increasing the bore is the most effective way to raise displacement because the cu. in. increase as the square of the bore — for example, a 1/8-in. overbore on the V8 increases displacement 8½%, while the same overstroke boosts it only 4%. That's why, dollar for dollar, you should bore before you stroke.

However, we didn't include boring in our "conservative" and "medium" souping categories because it is still not the most inexpensive way to raise HP. Boring your V8 block will cost you about $20, and with the oversize pistons and rings you'll need, your total investment will be $60-70, for an average HP increase of 10%. Not too good.

Now if you're souping only a little, and your block is badly worn and needs reboring, Ford supplies standard oversize pistons in several sizes up to 0.040 in. over at about $25 a set. However, at about this same price, there are several brands of inexpensive 4-ring road pistons on the market in sizes up to 1/8 and even 3/16 in. over; these should be considered if you must rebore anyway.

As for maximum bore recommendations on the Ford-Merc V8 block, experts disagree. Some late Merc blocks have been bored out as much as 1/4 in. and held up. But this was just luck, because block castings vary considerably due to "creep" and warp in the foundry. Thus when they bore originally in the factory, the bore may not come exactly in the center of the casting cylinder, and the wall may vary in thickness. Then if you get it too thin in spots by overboring, the wall will literally whip and chatter at high RPM and will quickly crack.

Therefore if you want your engine to stand up, we recommend a maximum overbore of 1/8 in. on the old pre-1939 Ford V8 block, and 3/16 in. on the later blocks. And we'd always recommend your going the maximum, because you might as well get the most for your money! Incidentally, some post-1939 blocks had very thin replaceable cylinder sleeves; these are no good for a souped engine and should be removed, and oversize pistons used.

Before going on, we might mention another boring practice that has been used on the V8 block — and that is boring out the entire cylinder wall (3-3/4 in.) and fitting "wet" sleeves. This is a terrific and expensive job, as it involves boring out the wall, counterboring for the sleeve flange, tapping into the sleeve for the head studs, fitting the sleeve, and then somehow getting the thing to hold water (on which they say the odds are about 50-50!) At any rate, you get a bore of 3½ in. by this method (since the sleeve wall is 1/8 in.), which would give 318 cu. in. with 4-1/8 stroke! The job would run well over $100 without the pistons, but that extra displacement could be the difference between success and failure for a track engine. It's a questionable practice at best, though, so investigate very thoroughly before you take the jump.

And then there is honing. Opinions vary on the necessity of this, but it's generally agreed that it's beneficial on very hot engines. And there are a good many ways of doing it — we know one guy who got beautiful, mirror-like cylinders with lard and toilet paper! The shop that bores your block will probably have some policy on this.

THE CRANKSHAFT

The crank is the heart of your engine. And it's got to be set up right if it's going to pump 200 hp at 4500 rpm for very long! But give it half a chance, and the toughness of this piece of stock equipment will amaze you.

If you're in the "conservative" and "medium" souping categories, the crank will be no problem. As we mentioned earlier, just "mike" the main and rod journals at several points to see that they're not worn over 0.001 in. undersize. (Standard main journals are 1.999 in. on pre-1937 blocks, 2.399 in. 1937-1938, and 2.499 on all later Fords and Mercs; rod journals are 1.999 in. on all pre-war Fords, and 2.139 on all Mercs and post-war Fords.) If your crank is worn too much, you can have it ground to take a standard undersize bearing; this job will run about $10.

But there might be a better way: If you have a 1939 block or later, you can use a standard late Merc crank (post-1949) without any changes! This has a 4-in. stroke, costs about $45, and should increase your HP some 7%. We've recommended this as a straight souping step on the hot road engine. (Of course, due to the long stroke, you'll also have to use

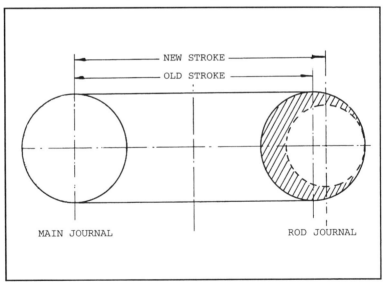

Fig. 5-2. Drawing illustrating what is done in "stroking"; the shaded area is ground off to increase the stroke length.

late Merc pistons, if you haven't bored, but the total price is still not a bad deal.) In other words, rather than spend a lot of dough to have an old 3-3/4 in. crank fixed up, why not put the money in a new 4-in. Merc crank?

As a matter of fact, the introduction of the long-stroke Merc in 1949 has thrown an entirely new light on this whole business of "stroking." You all know what this means — that is, grinding down the crankpins off center to lengthen the stroke; the stroke is lengthened by twice the off-set of the pin center and the diameter of the pin is reduced by the same amount (see Fig. 5-2).

Now with the possibility of using a stock 4-in. Merc crank, why lay out $25 to stroke an older crank to 3-7/8 in.? As we mentioned, a new Merc crank and a set of corresponding pistons costs about $70 — whereas a 1/8-in. stroke job on an old crank, new pistons, — AND NEW RODS, would cost you considerably more. And you would get less HP.

So we've just got to revise our views a bit here. The best idea on the usual souping job now is to forget all about stroking (on post-1939 blocks). Just drop in a late Merc crank and let it go at that. If you still insist on more stroke, you can stroke the Merc 1/8 in. (to 4-1/8 in.) and use special 3/8-in. stroker pistons and pre-war Ford rods (usually 21A type). Incidentally, all that bother and expense will boost your HP about 3%!

In view of all these things, we'd make the following crankshaft recommendations: In the "conservative" and "medium" categories, leave the crank as is, unless it needs a major regrinding. In this case, toss it out and use a new Merc crank — and since you've got the engine down, bore out 3/16 in. and fit special pistons. In the case of a hot road engine, use the new Merc crank anyway, perhaps stroking it 1/8th. For competition engines, when you're after that last HP, the extra stroke is important.

This brings us to that questionable trick of "metallizing." In this, they grind the crankpin down off-center, spray a heavy coating of molten metal around the pin (where it fuses like a weld), and then regrind to size. The idea is to greatly lengthen the stroke without decreasing the diameter of the pin, so they can use stock rods and bearings. Stroke increases up to ½ in. can be had in this way. However, it's a practice that has been much criticized because, quite often sections of the sprayed-on metal break off at high RPM — and really go to town in the lower half. In our opinion this science of metallizing is far from being developed yet, and since the late Merc crank gives a pretty substantial stroke to begin with, we suggest you forget about it.

That about takes care of the crankshaft. However, before we go on, we might mention some refinements for the guy who's building up a red-hot competition job, and nothing is too much trouble. One practice is to grind excess stock off the crank arms at the very "beefy" sections where it won't effect the strength in any way. About 6 lbs. of steel can be usefully removed in this way, and it will help your acceleration just a hair.

Another thing you can do is to machine and polish the shaft all over

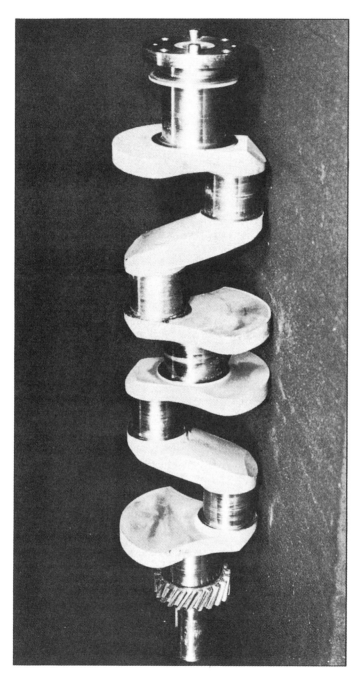

Fig. 5-3. Barney Navarro's special 180° crank with 3-in. stroke, built for his Class A Lakes car (176 cu. in.). These shafts are considerably lighter than the stock unit and increase acceleration somewhat, but have no effect on peak HP.

in an effort to reduce "oil drag." Oil will cling to a rough surface, which means its weight must be accelerated if the surface is in motion. By smoothing and polishing the crank surface, the drag of this oil is greatly reduced and there is a slight torque gain (we don't have any dynamometer figures showing the difference in HP between a rough and polished crankshaft, but the idea is logical anyway!).

One more thing—about these special 180° cranks. With these, two cylinders on the same bank fire simultaneously and we get, in effect, a four-cylinder firing order. The only advantage here is that the crank can be considerably lighter for added acceleration. However, because of the odd drawing impulses through conventional multiple manifolds, mixture distribution is poor, which kills off some of the potential acceleration. These cranks would only be practical for track work, but we won't take a stand one way or the other.

PISTONS AND RINGS

Pistons are a big factor in the performance of an engine. They have

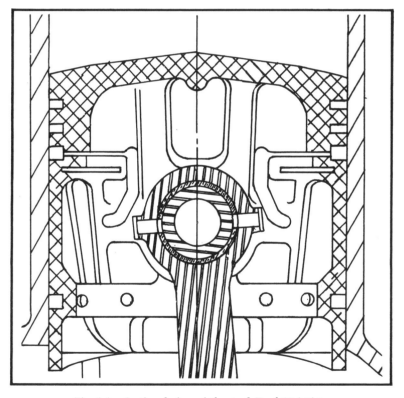

Fig. 5-4. Sectional view of the stock Ford 59A piston.

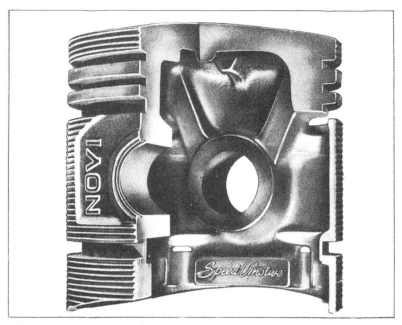

Fig. 5-5. *Cutaway view of a Speedomotive 4-ring road piston, typical of the more inexpensive units designed on the lines of the stock V8 piston and available up to 3/16 in. oversize. This type is best for any extended road use.*

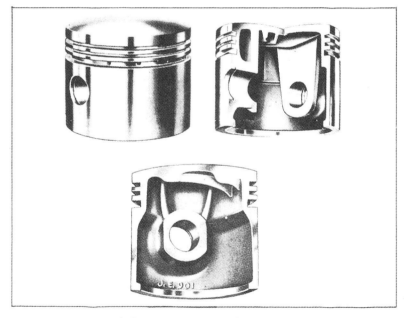

Fig. 5-6. *Cutaways of the JE racing pistons for the Ford-Merc, typical of the light 3-ring type. These are terrific for the track, but they don't stand up on the road.*

an exacting job to do, and they have some pretty horrible working conditions! But apparently it's a common fault among rodders to regard their piston problem with too much concern.

We refer to the practice of throwing special 3-ring racing pistons into any road engine with even a mild degree of "soup." A guy we know has had his set in 5,000 miles and you can hear him coming a mile away! The idea is this: Those 3-ring pistons were designed especially for competition where the engine will be torn down and gone over every two or three races — and not for 50,000 miles on the road. They figure that a 3-ring piston has 3/4ths the friction of a 4-ringer, and this is substantially true; also these racing pistons are set up with excessive clearances to further decrease friction.

The overall result is that the side-thrust on the piston due to the gas pressure on the head tends to rock it in the cylinder, which wears the rings at the top and the skirt at the bottom in a hurry. Oil consumption shoots up and compression pressure drops — so that often within 3,000 miles, you find yourself with a roaring case of "piston slap," a shot set of rings, and a not-so-hot engine. These light, solid-skirt 3-ring pistons are wonderful for racing when you can keep new rings going, and we can well imagine that they could make a difference of a good 5% in HP, but they're just not practical for a road engine.

Fortunately, several companies supply oversize 4-ring aluminum pistons for various bores and strokes on the V8 block, running anywhere from $20 to $45 a set; these are split-skirt and are built along the lines of the stock piston. They've given very good results on the road. As for special racing pistons, a number of companies supply a full line for every bore and stroke, at something around $45-55 a set. These are a bit lighter than stock, have solid skirts for maximum strength, and are a "must" for peak power output.

So we'd make the following piston recommendations: For the "conservative" and "medium" souping categories, stick to the stock pistons unless you're boring out over 0.040 in. For any road engine bored more than this, and this would include our 'hot" road engine, use a set of 4-ring split-skirt types and fit with about 0.004 in. clearance at the skirt. (Remember, the more clearance, the faster the wear.) For any competition engine that won't be used for extended transportation on the road, use the 3-ring solid-skirt type with about 0.008 in. clearance.

Rings call for no particular comment here except this: Use the highest-quality rings, follow the piston manufacturer's recommendations as to setup, and the ring-maker's end gap limits. For road engines, it is recommended that you use chromed top rings for maximum wear under hard driving conditions.

RODS

There's not much we can do with the rods in our souping work that will make them do their job better. Stock rods are plenty good enough, though it is suggested that you work with a brand new set if you're building up a competition engine (they run about $25 a set). Naturally, if

PISTON DISPLACEMENT WITH VARIOUS BORES & STROKES		
Engine	Increase	Cubic Inches
Ford V-8 85 (3-1/16" x 3-3/4" stock)	Stock 1/8 B	221 239
Ford-Mercury 95, 100 (3-3/16" x 3-3/4" stock)	Stock 1/8 B 3/16 B 1/8 S 1/4 S (Metallized) 1/8 B, 1/8 S 3/16 B, 1/8 S 3/16 B, 1/4 S	239 259 268 248 256 267 277 286
Mercury 110 (Post-1949) (3-3/16" x 4" stock)	Stock 1/8 B 3/16 B 1/8 S 1/8 B, 1/8 S 3/16 B, 1/8 S 3/16 B, 1/4 S (Met.) 5/16 B (sleeve), 1/4 S	255 276 286 264 285 296 305 328

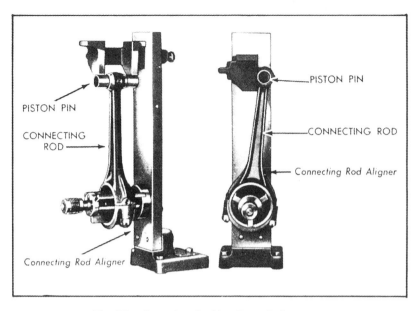

Fig. 5-7. Setup for checking V8 rod alignment.

you have the engine down, all rods should be checked for alignment at a Ford service shop (see Fig. 5-7). One thing you can do on rods for a competition engine is to polish them to decrease oil drag, as we mentioned earlier concerning the crankshaft. We don't know how much good this does, but it's worth considering. When assembling, rod nuts should be tightened to 40 lb-ft. torque with conventional nuts, and 45 with self-locking nuts.

BEARINGS

The job that the bearings have to handle in a stock engine is tough enough — double the HP of that same engine, and you've got trouble! And yet many hot-rodders pay little attention to their bearings and lubrication system. It used to be 30 years ago that, if your plain bearings wouldn't do the job, you would put in ball or roller bearings and bathe the whole kaboodle in castor oil!

With today's vast improvements, tests show that the friction of a shaft running in a well-lubricated plain bearing is no greater than steel balls rolling on a steel race — and the load capacity is a good deal greater. So you soupers needn't pine for a ball-bearing lower end. But also don't expect too much of your stock bearing setup.

The principle of a lubricated plain bearing is quite simple: Oil is fed in under pressure through a hole located at the point of least loading. The rotating shaft picks up the oil by friction and carries it along; as it approaches the point of maximum loading, the oil is "wedged" between the shaft and bearing, keeping them completely separated by a thin film of lubricant. The shaft is made hard and the bearing is made quite soft so that any metallic contact in the bearing (there is always a little) will rub off the soft side and imbed the metal particles without causing the bearing temperature to shoot up.

That's simple enough. But there's a catch: To get the desirable long-wearing characteristics with stock bearings, very soft metals like lead, tin, and copper are used. So when you double your average bearing loads as you do in a well-souped engine, the extra pounding wears them down in a hurry — or sometimes flakes them to pieces or burns them right up!

The solution is obviously harder bearing linings. A popular lining for this purpose is a cadmium-base metal (97%), with small amounts of either nickel or silver alloyed. These "cadmium-nickel" (or silver) linings are available as replacements for the V8 engine at around $17 a set (rods and mains), and they are giving excellent results — though don't expect them to stay 100,000 miles in a hot engine!

So here are our bearing recommendations: The camshaft bearings are lightly loaded, and the stock linings will be okay in all cases. For the "conservative" and "medium" categories, rod and main loads are not too bad either, and stock bearings clear through should be okay. On hot road and track engines, the crankpin loadings are considerably higher and harder linings than stock should be used here, though stock mains will be okay. For the "super-competition" engine, cadmium-silver linings on rods and mains are needed. Incidentally, never use hard rod bearings if your crank has been metallized.

Here's a tip: Don't use the late post-1949 doweled rod bearings for the heavier souping jobs. Get a set of pre-1949 rods and floating bearings—they will stand up a lot better.

Now a word about an important matter — bearing clearances. Obviously, by increasing this, our bearing will flow more oil, the film wedge will be thicker, and the bearing will be able to carry a higher load without severe metallic contact (this is only true up to a point). For the Ford-Merc V8 block, stock rod and main clearances should be roughly 0.002-0.003 in. This is okay for mild souping, but for the hot road engines, these clearances can be increased 0.001 in. — and even 0.002 on the super-competition jobs. You'll get some healthy pounding, but that's how it is in this souping business!

LUBRICATION

We don't expect you to be concocting any secret lubricant formulas like Leadfoot Louie's idea of a 50-50 Gulfpride-drain oil mixture — but for heavens' sake, don't try to run anything at 30 lbs. pressure on a track job like he did!

We discussed general lubrication theory in connection with the bearing problem, and we learned that a film of oil was supposed to completely separate the bearing surfaces at all times. Under otherwise similar conditions, the bearing load that this oil film can take without collapsing depends on the "viscosity" of the lubricant and the feed pressure from the pump.

Viscosity refers to the "flow thickness" of a fluid — for example, an oil of high viscosity requires longer for a certain amount to drain through a certain orifice than one of "lighter" or lower viscosity. The higher the viscosity, the thicker the oil and the more load the film can support.

But there's also the matter of oil pressure. The higher this pressure, the greater the flow rate and the thicker will be the oil film under a given load. The overall result is that we can increase the load capacity of our bearings "artificially" by increasing oil viscosity and feed pressure. But on the other hand, we must remember that thicker oil works against the feed pressure from the standpoint of flow rate. So we have to work out a compromise in our souping something like this:

Fortunately, it's not too tough to boost your oil pressure, since the Ford lubrication systems are fitted with relief valves and the pumps are capable of considerably more pressure than their rating.

On pre-1949 engines, a smaller pump was used for a normal oil pressure of 30 lbs., and the spring-loaded relief valve is located under the valve chamber cover at the front of the block. By replacing the spring here with a late (post-1949) high-pressure spring, the valve is held on the seat with more force and she'll push 50-60 lbs.

On the late V8's, the whole oil pump has been redesigned with larger gears giving considerably greater flow capacity, and with the relief valve in the pump body; normal pressure is 45-60 lbs. For 1950, the pump gears were changed from spur to helical teeth to provide even greater flow capacity, and will pump up to 80 lbs. under favorable conditions with No. 30 or heavier oil (and because of the new tooth design on the post-1950 pumps, these are preferable to the '49 for super-competition engines).

Another possibility for boosting your pump output is to merely stretch the spring out, which can give pressures well over 100 lbs. This is a guesswork deal at best, and should be regarded with caution. We know a guy who yanked his spring out a mile, went raring around with 120 lbs. oil pressure — and burned up a good set of rod bearings in no time! Such terrific pressures literally "sand blast" your bearings and accomplish nothing. Never go above 80 lbs. Spring-stretching is okay, but don't overdo it.

So we'd recommend the following in regard to your oil pump: Since all the pumps are interchangeable in the different blocks, we'd suggest you fit a 1950 or later pump and run 80 lbs. pressure for any track competition engine. For a hot road engine, a late pump is advisable too, but 60 lbs. pressure should be enough. On the milder souping jobs, the stock setup will be okay.

Now how about the different oils? The day has passed when you need castor to keep an engine in one piece at 4000-5000 rpm. We have some wonderful mineral oils available to us today that can do castor's job with a lot less fuss and bother. As a result, since castor is a vegetable oil — which means it will oxidize to gums and corrode engine parts — it is seldom used anymore, especially since it costs more too.

Therefore, keeping in mind all our requirements as to oil film strength, cooling, oil flow, and friction, we'd suggest the following S.A.E. oil grade viscosity ranges for our various souping categories (assuming summer conditions): For the "conservative" and "medium" classes, just conventional SAE-30 should do, about as stock. For a hot road engine, driven pretty hard, No. 40. For a competition job, running fairly long races, No. 50; for very short sprints, No. 20 or 30 will reduce friction, but watch your oil temperature!

Now there are certain expensive brands of oil additives on the market that are considerably better as regards film strength and friction than regular oils.

Though their individual refining processes are secret, the general principle here is a chemical ingredient that gives the base oil the "oiliness" and affinity for metal of a vegetable oil, but without the harmful gumming and corroson. And they really work, too; dynamometer tests on one highly-souped stock engine (over 200 hp) showed a peak HP increase of 5% by just switching to Wynn oil! We hate to come out and say to use it or don't use it — but consider the question thoroughly, at least for for competition engines.

Incidentally, an oil temperature gauge is a good investment for your souped engine. Fig. 5-8 shows how quickly the viscosity of an oil falls off with increased temperature. Obviously this is sure death to the lower end if it goes too far. A good running temp is 170-180° F.; don't exceed 200 in track racing under any conditions, even if you have to fit an oil cooler. (Actually, oil cooling should not be a problem except perhaps on the hottest competition types.)

THE FLY WHEEL

The weight of a well-balanced flywheel can have absolutely no effect on your HP at any RPM! Then why should we lighten it? It's all a matter of acceleration. Just as it requires a FORCE to accelerate a body in a straight line, so it requires a TORQUE to accelerate it in rotation. Just the extra torque in low gear required to accelerate the rotating parts of your car's drive line (crankshaft, flywheel, drive shaft, wheels, etc.) is equivalent to adding some 40% to the car's total weight — and the flywheel alone accounts for nearly half of this!!! No wonder we want to chop it — even though the idling and very low-speed lugging suffers.

Our stock Ford-Merc V8 flywheel weighs 39 lbs. There are a lot of shops that will turn it down on a lathe to 18-24 lbs. without weakening it in any way, for maybe $10. Besides this, special aluminum wheels (12 lbs.) are available from a number of companies for $25-30. The aluminum jobs, however, don't stand up too well under the usual day-to-day clutch pounding of a road engine, so we recommend that they be used only for racing.

BALANCING

Here's something that a lot of rodders forget altogether — and a lot of them get in a tizzy over nothing on balancing. Here are the facts:

The dynamic forces at work in an engine at 5000 rpm are something

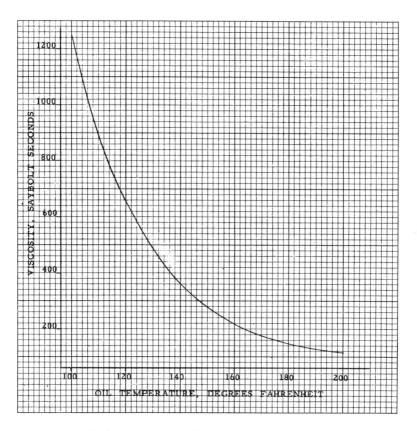

Fig. 5-8. Graph showing how quickly the viscosity of an oil drops off with increased oil temp; this is for SAE #50 grade.

scandalous! There are "reciprocating inertia" forces pulling up and down on the crankpin caused from starting and stopping the pistons and rods twice each revolution (see Fig. 5-9). Then there's the centrifugal force of the lower end of the rod spinning around the crankpin. And there's the centrifugal force of the crankshaft throws spinning around the main journals. Add to this the gas pressure pounding on the piston heads — and you've got one terrific balancing job on your hands!

The constant centrifugal forces, of course, can be readily balanced with counterweights, but on a 4-throw crank, the best we can do with the sum of the gas pressure and reciprocating forces is to balance ½ of them with counterweighting (the reason is complicated). But here's the point: When we juggle those rotating and reciprocating weights by using oversize pistons, non-standard rods, by removing stock when stroking, etc., we foul up the original counterbalancing.

In this case, the crank assembly should definitely be balanced. But

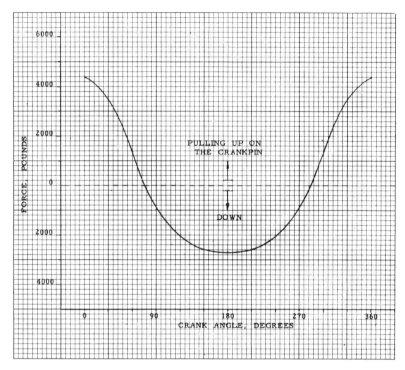

Fig. 5-9. Reciprocating inertia forces exerted on the crankpin by the weight of the piston and rod in a typical stock engine at 5000 rpm.

when you haven't made any great alterations, as on the "conservative" and "medium" engines, we see no need for rushing to the balancing shop, especially since the job will run about $35 and won't add a HP. Incidentally, any flywheel that has been chopped should be rebalanced.

COOLING THE V8

The V8 has always been difficult to cool. The stock setup will barely handle the situation when the block is putting out 100 hp — double that, and it's a battle of the B.T.U.'s all the way! We could hardly put ice water to 'er like the "Lakes boys" — and certainly Leadfoot Louie's idea of ripping the radiator and grill out and air-cooling it is questionable — so we'll just have to seek our solution on the engine itself.

Souping men have been experimenting with the problem for 15 years, and now they have it pretty well licked. Ed Iskenderian (the cam man) says that 90% of the cooling trouble with the V8 is caused by hot cylinder gas under compression or combustion escaping past the head gasket into the water jacket. The solution here is to coat both sides of the gasket

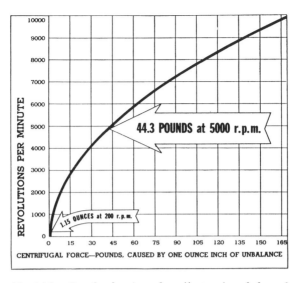

Fig. 5-10. Graph showing the effects of unbalanced centrifugal force in the crank train.

Fig. 5-11. Typical electronic crank balancing setup. Bob-weights on the crankpin are used to allow for the reciprocating and rotating forces of the pistons and rods, and electronic equipment indicates the unbalance when the crank is spun by the belt drive; metal is ground or drilled off the crank arms to bring it into perfect static and dynamic balance. Price, about $35.

with RED AVIATION PERMATEX gasket cement before final assembly.

Besides this, there are several other tricks that Iskenderian suggests to assure ample cooling under the most extreme conditions. The stock water pumps on the late V8's churn and whip the water too much to give full flow at high speed; a suggested cure is to cut off every other impeller blade and then drill a 1/4-in. hole through the center of each of the three remaining blades.

Also, you should either leave a thermostat in the head water outlet; or if this fails, install a "restrictor" to slow down the flow through the block and hold it under a slight pressure at all points. As a restrictor, use a metal disk with a locating sleeve welded on, extending down into the head outlet and locked by a pin through the neck and sleeve, retained by the hose. (Some guys just stick the disk in and depend on the hose to retain it in place.) Iskenderian recommends a 5/8-in. hole in the disk; others have been successful with 1/4-1/2 in. restrictors. As a further precaution, he suggests plugging the large water hole in the center of the block face.

Finally, Ed suggests that the best setup of all is to use an early water pump unit ('32-'36), mount it low in front of the engine at pan level, and drive it by belt at about ½ crank speed; use a "Y" branch off the regular outlet to pipe the water to the bottom of the water jackets on each bank. Along this same line, Martas supplies a special double-outlet water pump for the V8, mounted on the two forward pan bolts and driven directly by the crank (see Fig. 5-12); the pump feeds through two water manifolds to several points on the base of each bank.

At any rate, you should plan some mild cooling precautions as out-

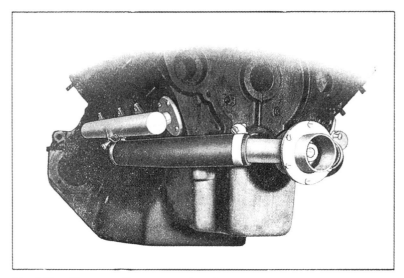

Fig. 5-12. Martas cooling system for the V8, maintaining pressure at the bottom of each bank.

lined earlier for the lighter souping jobs, and perhaps something like the Martas or single Ford pump for extreme competition conditions.

CHROMING

In concluding this chapter on fitting up the block, we might mention one refinement in the souper's art of block work that is gaining in popularity all the time — chroming. Actually, there is some controversy as to the need for this business, and we think they are going too far with it in many cases. The only real purpose of chrome plating such parts as the crank journals and cylinder bores is to make them so hard that they'll wear practically indefinitely (the slight reduction in friction couldn't make too much difference).

This long-wearing is wonderful, of course, but we must weigh the advantages against the costs. Chroming is quite expensive, running some $70 to do the rod and main journals on a V8 crank. If you are trying to preserve a $250 Offy crankshaft, chroming is the way to do it — but we wonder if it's worth it on inexpensive stock parts. We won't take a firm stand on this because a lot of things enter into it, but, in general, we wouldn't worry too much about chroming on a stock block.

That about concludes our consideration of the block. We'd like to repeat again here that your block setup is vital to the success of your engine. Don't slight anything. But there are also plenty of places to throw away money here too, so be conservative. Weigh every move in the balance of advantage vs. cost.

Now let's swing along to another branch of the "horsepower tree." Let's take a look at the cylinder head.

Nicknamed "The Bug", this short-chassis sprint car consistently ran through 1/4 mile traps from start at over 120 mph. For the short run, the two tubes carry water in place of radiator. Mercury power unit displaces 274 cu. in.

Combination road-and-drag roadster has big Mercury engine fitted with Evans high-compression heads. Note that longitudinal leaf springs replace stock Ford springs.

CHAPTER 6

CYLINDER HEADS

YOUNG Ernest Henri —burning the midnight oil over a Peugeot drawing board, designing for the 1912 French Grand Prix — was sure he had at last found the perfect cylinder head for racing. Under his flying pencil, the double-overhead-cam, dual-valve layout was born. And let's face it, Mr. Henri DID HAVE THE PERFECT HEAD!

But don't let Leadfoot Louie hear us. Certainly no little Frenchman forty years ago could outrun him — if we'll just let him whack an eighth off his Ford flats! Louie's favorite cure-all is the cylinder head; if it won't go as he thinks it should (and it never does), the first thing he thinks of is the head. We remember the time he shaved 0.150 in. off some 9:1 Edelbrocks!

Anyway, this problem of cylinder heads in our souping is both interesting and vital, principally because the stock engineers have given us so little to work with. The Ford-Merc is no exception. We well realize that 99 out of 100 of you rodders can never hope to go beyond flat heads in your V8 souping — and we have been criticized for blowing so much about overheads — but we will stick to our guns.

Here's why: Properly-designed overhead valves have a HP margin of a good 20% over side valves. That is a simple and straight fact. In other words, if you sweat it out on a flat-head V8 block and finally come out with 230 hp, then you can turn around and get about 230 x 1.20 = 276 hp, by just switching to overheads!

Let no man sneeze at this. A lot of hot-rodders are prone to belittle overhead valves on the V8 block as a lot of unnecessary complication and expense. Admittedly they are expensive and complicated — and aren't really practical for most rodders — but when you're after that last HP, they are as necessary as a runway in a burlesque house! This matter of overheads and L-heads is the very basis of our head souping problem, so it might be a good idea to investigate all angles briefly:

OVERHEAD VS. SIDE VALVES

In the first place, the advantage of one head layout over another from a strict HP standpoint does not lie in the SHAPE of the combustion chamber. This has its effect, of course, but the more important factors here are: (1) The intake gas flow path from the valve into the cylinder, and (2) the space available for the valves. Fig. 6-1 compares overhead valves and an L-head from these two standpoints.

Consider first the gas flow. As we learned earlier, our basic souping aim is to get the maximum weight of fuel-air mixture into the cylinder on the intake stroke. Also we learned that, whereas the cylinder will draw in a volume of gas equal to its actual displacement in cu. in., the true weight taken in will depend on pressure and temperature as well as volume.

Our enemies here are turbulence within the gas and friction between adjacent layers of the gas and between the gas and channel wall. A great deal of the density loss in the intake gas comes after it has passed through the valve. Notice in Fig. 6-1 that, with the L-head, there is a lot of turbulence and friction of the incoming gas against the combustion cham-

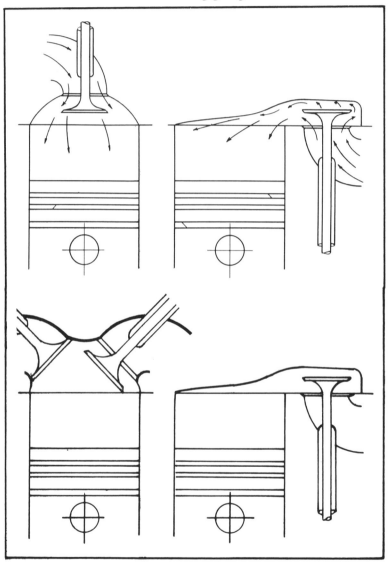

Fig. 6-1. WHY OVERHEAD VALVES ARE SUPERIOR TO SIDE VALVES. Top—Smoother, non-turbulent gas flow into the cylinder. Bottom—By inclining the valves at an angle to the cylinder axis, much larger valves can be used. With side valves, diameter is limited to about $\frac{1}{2}$ the bore; if we incline the valves in the head, we can increase valve diameter 30 or 40%.

ber surface; this all acts to expand the gas and decrease the weight per cu. in. On the other hand, with overhead valves, the flow into the cylinder is very smooth and easy. Tests on one engine showed a volumetric efficiency increase of 16% by just moving the valves from the block to the head!

And then there's the matter of valve size. Naturally for maximum "breathing," our valves want to be as large as possible and have a maximum of discharge area. With an L-head, the size of the valves is pretty much limited by the bore (see Fig. 6-1) whereas, with an overhead layout, we can incline the valves at an angle and increase valve diameter some 30-40% without too much trouble. This factor of valve size is vital when you're after maximum HP, as we'll discuss later.

So we see that the overhead-valve layout has it all over the L-head in the all-important matter of volumetric efficiency. Not only this, but with the compact combustion chamber with overhead valves, the pressure rise during combustion is faster and combustion is a bit more efficient. Also, you can get much higher compression ratios with the more compact chamber. So the cards are on the table — and let's not overlook overhead valves!

COMPRESSION RATIO

This brings us to another major factor in head design — compression ratio. You all know what this refers to; that is, the ratio of the total cylinder volume at bottom stroke to the volume at top stroke.

It's a fairly simple matter to measure your compression ratio. Just position the crank to top center on the cylinder you're working with, then pour oil from a measuring beaker into the spark plug hole till the oil rises to the bottom of the hole. You can then calculate your compression ratio from the following formula:

$$C.R. = \frac{C + D}{C}$$

where "C" is your combustion chamber volume, measured by the amount of oil poured from the beaker, and "D" is your cylinder displacement. (If you are working with a "c.c." beaker, remember that 16.4 cc = 1 cu. in.)

For example, suppose we have a late Merc bored 1/8 in. — that's 34.5 cu. in. displacement per cylinder — and we measure our combustion chamber volume as 75 cc. First we convert cc to cu. in.; 75/16.4 = 4.57 cu. in. Therefore our compression ratio is:

$$C.R. = \frac{4.57 + 34.5}{4.57} = 8.55 \text{ to } 1$$

Now just what effect does compression ratio have on HP? It's obvious that it would increase the HP, other factors equal, because it would raise the compression pressure at the time the spark fires — and this would boost the effective cylinder pressure throughout the power stroke.

For example, say our engine has 6½:1 compression ratio and has a compression pressure of 150 lbs./sq. in. at 2000 rpm, with a peak cylin-

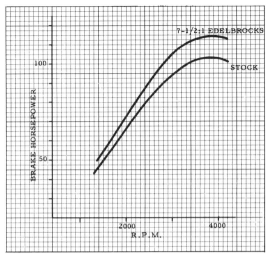

Fig. 6-2. Effect of high-compression aluminum flat heads on V8 performance (no other changes).

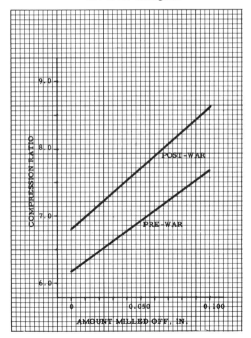

Fig. 6-3. Approximate compression ratio obtained by milling stock heads.

der pressure during combustion of 600 lbs. Now if we just raise that compression ratio to 10:1 without any other changes, our compression pressure jumps to 260 lbs., peak pressure to about 1,000 lbs., and — what's most important — the average cylinder pressure on the power stroke (which is equivalent to HP) jumps 15%!

This is the immediate, and all-important, effect of raising compression ratio. Fig. 6-2 gives power curves for an actual dynamometer test showing the effect of increased C.R. on HP over the full RPM range. Notice that this, like increased piston displacement, is quite effective in boosting HP at low speeds. This is important to remember if you're designing a hot street engine.

And there is another advantage in high compression too, though you may not be much interested from a souping angle, and that is improved gas mileage. Since we squeeze more work out of every drop of fuel by compressing it to a higher pressure before firing, our specific fuel consumption drops. Raising compression to 10:1 on the V8 will increase fuel mileage some 11% under similar driving conditions!

Now, lest we get the idea that higher compression ratios are all "gravy," let's remember the disadvantages. For one thing, our C.R. is sharply limited by fuel "knock," and we can't go much above 8½ or 9:1 on pump gas without running into trouble (we'll discuss this later). And there is also a little matter of engine wear. The extra pounding caused by the high peak pressures is hard on rings and bearings, even assuming you pull the same torque from the engine. We'd say you'll do well to get 30,000 miles out of a set of bearings with a 9:1 road job.

So high compression has its advantages and drawbacks — but the very fact that it will boost HP at all speeds is excuse enough for going "all-out" on it! Now just what can we do?

REWORKING STOCK HEADS

To begin with, let's not kid ourselves into believing that all the magic in the Arabian Nights will ever make a reworked stock head perform like a special speed head. It can't be done. All the late stock heads are cast iron with combustion chambers contoured for maximum gas flow and combustion efficiency at 6.8:1 compression ratio. Special heads are cast in aluminum alloy, with huge water spaces for maximum cooling, and combustion chambers designed for 7½ to 10:1 compressions. You can't just chop 0.100 in. off a stocker and expect it to outpull an Edelbrock!

But if you're just fooling around with your souping, and don't want to sink $70 in a set of specials, you can improve your stock heads by "milling." This refers to the practice of machining off some of the bottom side of the head to decrease the volume of the combustion chamber and increase the compression ratio. Many shops will do the job for about $10. Fig. 6-3 shows the approximate compression ratio you'll get by milling off the pre-war low-compression heads and the late 6.8:1 jobs (this graph is just a rough guide and we can't vouch for it.)

Here are your milling recommendations: On the old pre-1937 engines with flat pistons, you can mill off about 0.090 in. without worrying about clearance over the valves — which, incidentally, should be 0.040 in. If you "fly-cut" into the head for the valves, you can take off up to 0.125 in. without badly weakening the sections. For late heads with dome pistons, you can go up to 0.060 in. without any changes; 0.090 by redoming the head to give the necessary 0.050 in. piston clearance, and up to 0.125 in. if you redome and fly-cut for the valves. Remember, piston and valve clearance under the head (with gasket in place) should be 0.050 and 0.040 in. respectively; gasket thickness of 0.055 in., and you can use modeling clay to measure (since castings often vary considerably, you had better check in any case).

Now there are also several methods of "filling" the combustion space to increase compression ratio, but this practice never gives good results and we're not going into it. One reason why this sort of thing doesn't work on the V8 are the critical cooling conditions with the cast iron heads. Filling just makes things worse. As far as that goes, you're very apt to have cooling trouble even with milled heads if you use the stock cooling setup, because the cast iron can't seem to pass the extra compression heat fast enough. "Hot spots" result, knock starts, and things go from bad to worse. For these reasons, we can't honestly recommend any kind of stock head reworking on the V8, if you're serious about your souping.

There are still a couple of other possibilities along this line that may offer better results. Ford supplies a special "Denver" head for high-altitude service with $7\frac{1}{2}:1$ compression; this has combustion chambers designed for this ratio and will give somewhat better performance than a standard 6.8:1 head milled. Also, for about $5 you can buy a set of special thin head gaskets that raise your effective C.R. to 7.13:1.

Incidentally, you are almost sure to develop a bad case of knock when you rework cast iron heads to $7\frac{1}{2}:1$ or better. This can be corrected to some extent by retarding the spark, as we'll discuss later in the chapter on ignition.

SPECIAL FLAT HEADS

You can feel mighty lucky that you don't have to depend on stock head reworking like Harry Miller did back in 1935. Today there's a host of special V8 flat-head equipment on the market, of every quality and every price, and ranging in C.R. from $7\frac{1}{2}$ to $9\frac{1}{2}:1$. Most of it is for the later 24-stud blocks, but there is still a fair selection of flats for the old pre-1939 21-stud jobs.

There are first-quality aluminum head sets for road and track work at about $75 a pair. Second-grade aluminum types range down to around $50 a set — and at the foot of the scale, for those who are in their souping for laughs, special cast iron high-compression heads are available down to about $25.

At these prices, special heads are a pretty fair investment from the standpoint of HP per dollar, and since they are quite effective in boost-

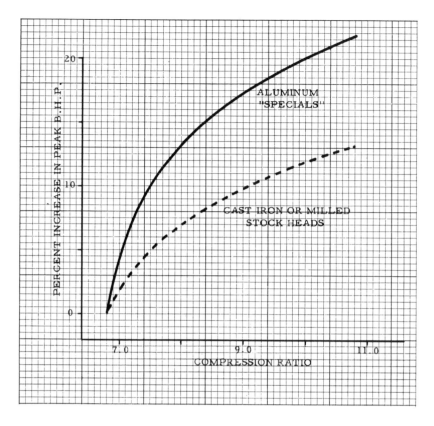

Fig. 6-4. Increase in peak HP obtained by increasing compression ratio (no other changes).

ing performance at low RPM, we recommend them for all our souping categories. Fig. 6-4 shows the approximate percent peak HP increase you get with different compression ratios on aluminum heads (with no other changes).

Fig. 6-5 shows the general layout of these special heads; notice the rugged construction, heavy sections, and ample water spaces. Aluminum in itself is vital for high-compression heads; it conducts heat 4.4 times faster than iron, and this lowers the surface temperature of the combustion chamber, increases volumetric efficiency somewhat, and allows higher compression on a given fuel without knock. Since cooling is critical on the V8 anyway, you need aluminum heads in your souping like you need your right arm — and for this reason, we can't recommend cast iron heads for any serious work. On the other hand, we see no reason

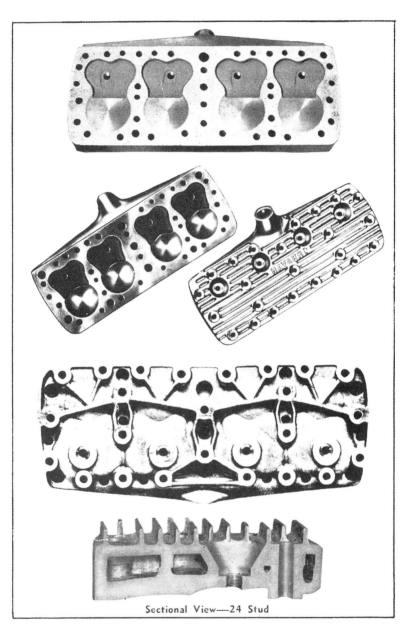

Fig. 6-5. *Several views showing construction features of typical aluminum heads for the V8, for stock domed or oversize pistons; the cutaways show the rugged internal construction and ample water spaces.*

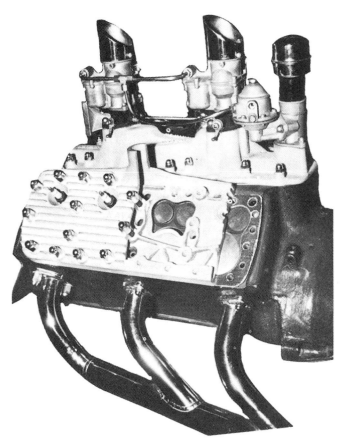

Fig. 6-6. Cutaway view of the Navarro head for domed pistons, showing combustion chamber, valve space, etc.

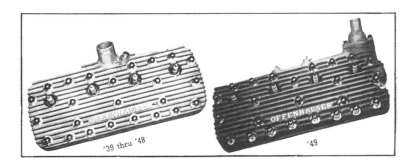

Fig. 6-7. Offenhauser 24-stud racing heads.

Fig. 6-8. Special Tattersfield heads supplemented with "pop-up" pistons extending into the combustion chamber for compressions up to 14:1. Strictly for racing.

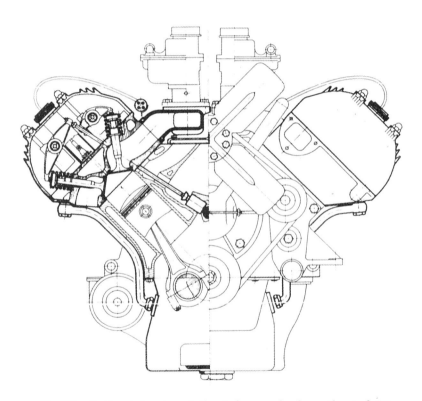

Fig. 6-9. Sectional drawing of the Ardun overheads on the stock V8 block; note the simple installation and ample porting. Various manifold combinations are available.

why you need top-quality aluminum heads for a road engine. Use your own judgment.

Incidentally, one thing we have to remember about our heads is this: Increasing the cylinder displacement by boring and stroking alters the ratio between cylinder and combustion-chamber volume, and increases the compression ratio. High-compression heads these days are rated on a basis of stock bore and stroke of the 239-cu. in. block (3-3/16 x 3-3/4 in.), so we must allow for any changes in cylinder displacement in determining our true C.R. (and also remember that relieving the block lowers the ratio a bit). The following table shows C.R. for the usual bores and strokes:

COMPRESSION RATIO CHART FOR THE "239" V8 BLOCK						
BORE & STROKE		RATED RATIO				
	STOCK HEADS	7.5	8.0	8.5	9.0	9.5
Stock bore & stroke	6.8	7.5	8.0	8.5	9.0	9.5
1/8 bore	7.3	8.1	8.6	9.2	9.7	10.2
3/16 bore	7.5	8.3	8.9	9.4	10.0	10.5
1/8 bore, 1/8 stroke	7.5	8.3	8.9	9.4	10.0	10.5
3/16 bore, 1/8 stroke	7.7	8.6	9.2	9.8	10.4	10.9
Late Merc crank, 4"	7.2	7.9	8.5	9.0	9.5	10.1
Above, with 1/8 bore	7.7	8.5	9.1	9.7	10.3	10.8
Above, with 3/16 bore	7.9	8.8	9.4	10.0	10.7	11.1

Now a word about recommended compression ratios. For road engines on pump gas, don't go above 8½ or 9:1. For competition engines on special fuels, 9 to 12:1 can be used. And then, of course, don't forget the special heads with "pop-up" pistons that extend up into the head to allow up to 14:1 compression without greatly restricting the flow path area into the cylinder, which is a severe limitation on conventional heads (see Fig. 6-8).

OVERHEADS FOR THE V8

There's gold in them thar' overheads — in dollars as well as HP! Fantastic outputs are available at a price, and it's a good idea for the guy who's in racing up to his neck to investigate the overheads very carefully. We know, of course, that few of you can lay out five or seven C's for this stuff, but let's look over the field anyway. Since the war, nearly a dozen different experimental overhead sets for the V8 have been designd and run with varying success, but as this is written, only four companies are actually producing equipment for public consumption — all for the post-1939 24-stud block. Here's a thumbnail sketch of each:

ARDUN — This head has been kicking around for several years and is now the most widely used of the V8 overheads. It is presently being produced by the Allard Motor Co. of London, and sells for around $500 complete. Fig. 6-9 shows the layout of the Ardun conversion; note the

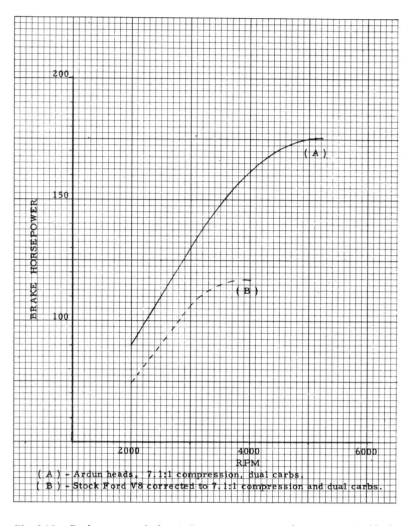

Fig. 6-10. Performance of the Ardun equipment on the 239 cu. in. block.

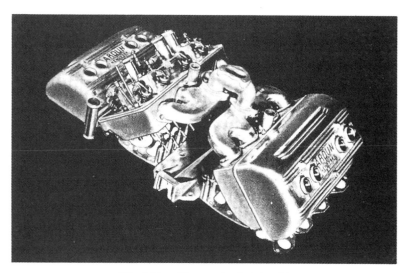

Fig. 6-11. The standard Ardun kit.

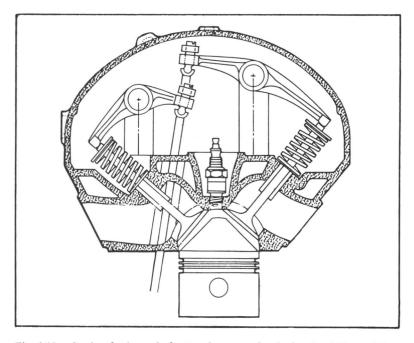

Fig. 6-12. Sectional view of the Stephens overheads for Ford-Merc. Using high-dome pistons with this type head layout allows terrific compressions up to 20:1 for racing.

huge valves and ports that allow 50% greater volumetric efficiency at high speeds. Fig. 6-10 shows a power curve for the standard Ardun conversion using dual carbs and 7.1:1 compression — the dotted curve is for the stock V8 corrected to dual carbs and the higher ratio. That terrific margin of HP is entirely due to the increased breathing! Full-race setups using this equipment should develop up to 290 hp at 5000 rpm on alcohol.

STEPHENS — This head, produced by the Stephens-Frenzel Co. of Denver, is becoming more popular all the time. It is designed especially for racing, whereas the Ardun was originally set up for marine and commercial installations. Figs. 6-12, 13 and 14 show the general layout. The head is cast iron, with full pressure oiling to the rocker arms, and employing stock Chevrolet valve guides and Cadillac springs. The latest design has an arrangement for shielding the spark plugs and wires from rocker oil. Breathing on the Stephens head is very good too, as evidenced by the power figures; on a '48 block bored 1/8th (259 cu. in.) with dual Stromberg "97" carbs and 10:1 compression, the peak is stated to be 185 hp at 4400 rpm on pump gas. Full-race setups have been run up to nearly 300 hp.

LEE "TORNADO" — This equipment has been under development for several years and is now in limited production; it was designed by Lee Chapel and is sold by Lee's Speed Shop of Oakland, Calif. Figs. 6-15 and 16 show the layout. The "Tornado" uses vertical in-line overhead valves operated by solid push rods, with four ports on each side of the head for intake and exhaust. The heads can be had for flat or domed pistons in compression ratios up to 16:1; the complete kit includes manifolds and special camshaft. They claim up to 250 hp in race trim with bored and stroked block.

Fig. 6-13. Stephens heads on the 59A block with close-coupled manifold for stock generator placement.

Fig. 6-14. Stephens-Ford engine in a hot roadster, with valve cover removed to show rocker arms and shafts. Something in the neighborhood of 1.0 hp per cu. in. should be obtained from this setup, under favorable conditions.

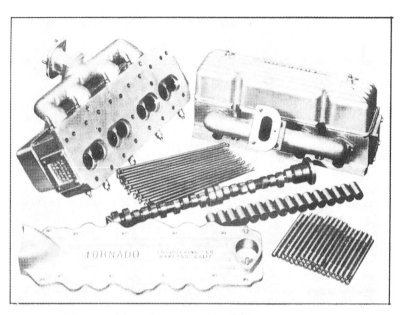

Fig. 6-15. The Lee "Tornado" V8 overhead kit, using a separate carb and manifold for each bank.

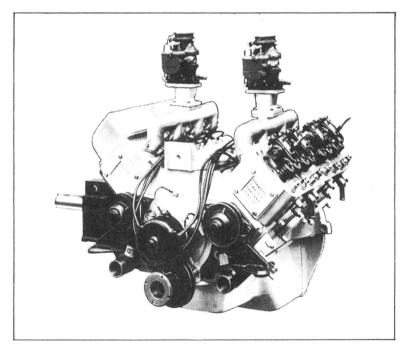

Fig. 6-16. View showing the Tornado equipment mounted on the 59A block.

Benson Ford, V.P. Lincoln-Mercury Division of Ford Co., looks over the Bob Estes Spl., with driver Joe James. The car, equipped with Ardun overhead valve setup, qualified as an alternate starter in the 1950 Indianapolis Race at the amazing speed of 124.176 mph for a semi-stock design.

We believe this design fails to capitalize on overhead valves' main advantage, and that is the possibility of inclining the valves to make room for very large diameters. By lining them up vertically, they can't use any larger valves than they could with an L-head (the "Tornado" intakes are 1.62 in. compared with 1.81 for the Stephens, and breathing suffers. It does make a much simpler and more economical layout, but it is a compromise.

SPEEDOMOTIVE—This head, sold by Speedomotive, of Covina, Calif., is a rocker-overhead outfit featuring 2-in. intake valves and compressions up to 18:1.

That about concludes our considerations of the head problem. It's a matter we must look at very carefully if we're going to squeeze that last HP out of a stock block — or even squeeze every possible HP out of our souping dollar. The cylinder head is certainly no cure-all, as Leadfoot Louie thinks. No one factor in engine design is.

But if there is one factor that comes close to it, it's the subject of our next chapter — the induction system.

CHAPTER 7

THE INDUCTION SYSTEM

LET'S face it — we're beating our heads against the wall on this induction system nightmare! If we could somehow solve the problem of nursing in a cylinderful of soup 40 times a second under Mother Nature's puny atmospheric pressure, 3/4ths of our souping worries would be over.

Harry Miller thought he had the answer 16 years ago with four duplex carbs. California thought it had the answer with wild cam grinds. Stuart Hilborn thought he had the answer by tossing out the carb and manifold and injecting the fuel right into the port. And now the "super-fuel" boys have thrown in the sponge on the induction system in general — and are carrying their oxygen around in a can!

Where it will all end, we don't know — probably Leadfoot Louie will get his fool head blown clear off — but we're definitely approaching a peak. Europe would tell us to slap on a supercharger and forget it, and we're inclined to think they're right. We'll go into this later, but right now there's an awful lot we can do with that stock induction system. In fact, no other souping step will pay off in more peak HP — and less low-RPM power — than this one. Let's start right at the fuel tank:

FUELS

Your fuel can have as much influence on HP as more spectacular things like heads, cams, carbs, etc. The subject should be investigated thoroughly if you're after that last HP for the track. We'll only review things here, since we went into considerable detail in our previous book, "Souping the Stock Engine."

(We are not going into the subject of "super fuels" here. As you know, these are merely explosive compounds that carry their own oxygen — which, of course, alters the entire problem from a fuel standpoint. It's not an established souping practice just yet, so we'll drop it now.)

Okay — if we only consider the conventional fuels that require air to burn them, we find the three main factors that govern fuel selection are: (1) Octane rating, (2) volatility or tendency to evaporate, and (3) latent heat of vaporization. Of course, with a road engine, we'll have to burn gasoline that's available at the pump, so there's no real choice here (but remember to keep compression ratio below 9:1 to prevent knock). For competition, however, we're not limited to pump gas, and we can select the fuel that will give us the best overall performance within our fuel cost limits.

Consider octane requirements. We won't go into the technical explanation of octane number, but suffice it to say that this value indicates a fuel's relative resistance to knock and pre-ignition. For example, premium pump gas rates 80-90 octane (depending on mixture richness); benzol

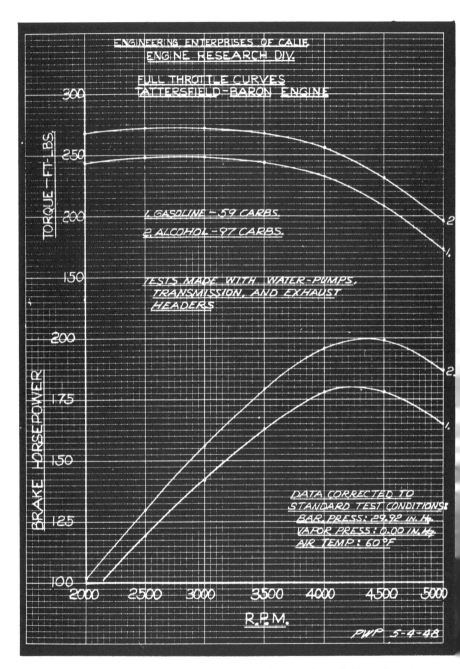

Fig. 7-1. Dynamometer test curves on a full-race Tattersfield-Ford engine showing 11% HP increase by just switching to alcohol fuel.

rates 87-120; alcohol, 90-200.

Now our octane requirements under high-output competition conditions in road or track racing with 9 to 12:1 compression — that is, the fuel octane we will need to prevent wrecking our engine from continuous knock, will range generally from 90 to 110. In other words, pump gas is not a good deal here. If cost is a big factor, 25-50% benzol mixed with pump gas will give the required knock resistance; otherwise 91-octane aviation gas (at around 40c a gallon) or the more expensive fuels will be okay.

A word of caution: Never use super-octane aviation fuels, such as 130-grade combat gas, in a souped stock engine; these burn very slowly and overheating is a sure thing.

Consider volatility. This isn't important in track racing where there is no low-RPM acceleration, but in road racing, we've got to consider it in order to get good mixture distribution at low speeds. Here is a list of the usual racing fuels in the order of their volatility:
1. Ether
2. Acetone
3. Gasoline
4. Methanol (wood alcohol)
5. Benzol
6. Ethanol (grain alcohol)

Because of this volatility, you can't very well use pure alcohol for road racing, but they often mix in 25-50% gasoline, or sometimes 5% or less of ether or acetone.

From the standpoint of pure HP, the most important characteristic of a fuel is its "latent heat of vaporization." As we all know, any liquid absorbs heat when it evaporates and expands to a gas. Let some alcohol evaporate from your hand and notice how cold the spot feels. Anyway, when the liquid fuel vaporizes in the manifold, it absorbs heat from the air in the mixture — which contracts it, so that we draw a greater WEIGHT of mixture into the cylinder on the intake stroke. And our HP shoots up in proportion!

Fuels vary widely in the amount of heat they absorb when evaporating, and this amount of heat, measured in B.T.U's per lb., is called the latent heat of vaporization. Here is a list of our fuels with their latent heat values:
Methanol, 500
Ethanol, 390
Acetone, 235
Benzol, 162
Ether, 160
Gasoline, 135

This is why alcohol will show a peak HP increase of about 10% over gasoline without any other changes (see Fig. 7-1). So from this standpoint, pure methanol is absolutely our best bet for maximum possible HP. The disadvantage, of course, is the very high fuel consumption with

Special chopped coupe is fitted with 270-cu. in. Merc, burns nitro mixture—"super fuel"—turns near 120 mph. in ¼ mile drag races in So. Calif.

Appropriately called "Ant Eater" by hot rod fans, this car is also in the 120-mph class, has a large-bore Merc engine, but has dual rear wheels for better "dig" from standing start.

alcohol. If we're running in a long-distance race, the pit stops for fuel might cost more in average speed than the 10% less HP with gasoline But for short sprints, you can't beat methanol.

So that covers the essentials of the fuel problem. Let's sum it up thus: If you're not going all-out with your racing and fuel cost is a big factor to you, a mixture of pump gas and benzol or 91-octane aviation gas will do. If you are serious about sprint competition, you can't beat pure methanol (not ethanol). For long-distance racing, just gasoline alone will give the maximum fuel tank mileage, but you'll have to weigh other factors like cost and HP to make a final decision — but probably 50% alcohol is plenty for this type of racing. For road racing, we need a fairly volatile fuel; gasoline is good in this respect, but if we mix more than about 60% alcohol, better figure on a small amount of ether.

FUNDAMENTALS OF GAS FLOW

Since the whole problem of the induction system from this point on is merely one of "gas flow," perhaps we should pause right here and study some of the basic fundamentals of this subject. Actually, our fuel-air mixture is not really a true gas, but is a gas (air) with atomized liquid in suspension, since not over 10% of the fuel evaporates to a gas outside the cylinder. However, for all practical purposes we can treat our fuel-air mixture as a true gas — and as such, it will have viscosity like an oil, friction, weight, momentum, etc.

Now the science of fluid dynamics teaches us that, when a gas flows through a closed channel, the gas pressure will be less at the outlet than the original pressure at the inlet. You can't get away from it. This pressure loss is due to two things: (1) Friction of the gas against the channel wall and friction between adjacent layers of the gas flowing at different velocities, and (2) inertia losses within the gas due to turbulence, surging, "wire-drawing," etc. Pressure losses from these causes increase AS THE SQUARE OF THE RATE OF FLOW!

Now a further consideration of our induction problem from the theoretical standpoint shows that we can reduce pressure losses by the following methods (assuming we don't sacrifice our total flow rate, which determines our HP):

1. Increase the total cross-sectional area of the channel
2. Smooth the channel walls
3. Improve the channel path; large-radius turns, no sharp corners, etc.
4. Shorten the flow path length
5. No obstructions in the flow path

This is a pretty big order to fill in our souping work, but it at least shows us what directions to take for improving things.

Here is an essential point: Since induction pressure losses increase as the square of the flow rate—in the case of an engine, flow rate would be approximately proportional to RPM — we find that gas flow refinements will help matters a lot at high speeds, but not much at low. In other words, knocking yourself out on your induction system will have very little effect on low-RPM performance — because losses with the stock setup are al-

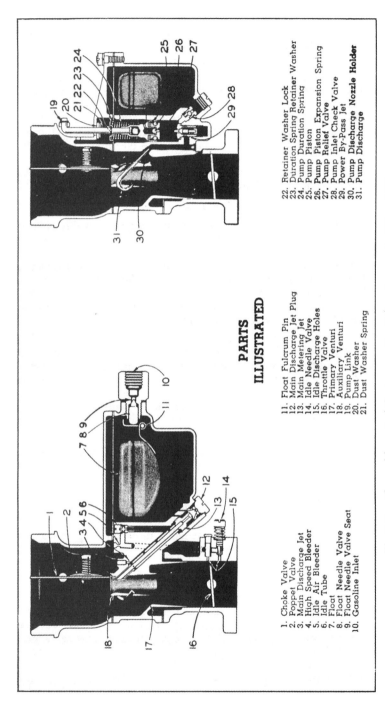

Fig. 7-2.: General layout of the pre-war Ford-Stromberg carbs (48 and 97).

ready very small below about 2500 rpm.

And here's another thing to remember in this connection : With the stock setup on the V8, approximately 75% of the TOTAL induction losses come between the manifold and the cylinder—in other words, downstream of the manifold. With dual carburetors, the proportion jumps to over 90% (because the flow rate in the manifold is halved, and losses are cut by 3/4ths), and with triple carbs, nearly all the losses are in the ports, valves, and cylinder.

This is not hard to appreciate when we consider the great flow restriction offered by the valve, its stem, guide, etc. This is all by way of pointing out the vital importance of block and valve modifications in reducing pressure losses. Don't just slap on dual or triple carbs and consider your gas flow work finished — it has only begun!

Now let's start right at the beginning of our gas flow path — that is, the carburetor — and go over our problem with a fine-tooth comb.

THE CARBURETOR

This is the "lung" of the engine, and it's the least of our souping worries. Stock carbs are right at home feeding red-hot setups, and in fact, there's very little we can do to increase their efficiency. Ford carbs are no exception in this respect, being used almost exclusively on soupd Ford-Merc V8's.

Except for the first 1932 and '33 models, all Fords prior to '37 used the duplex Stromberg EE-1 Model "48" carb, with 1.03-in. venturi and 0.048-in. main jets. In 1937 the Stromberg Model "97" was adopted, with the same layout, but using 0.97-in. venturis and 0.045 jets. Fig. 7-2 shows the layout of these carbs. In 1938, Ford began sponsoring the production of their own carbs through Chandler-Groves (Holley) and succeeding models were known as "91-99" and "59A" for the post-war engine; these carbs are similar to the old Strombergs, but there is a removable nozzle bar and the main jets are "buried" in the fuel bowl. Venturi diameter here is 0.94 in. and 0.050-in. main jets are standard. For 1949 and later engines, Mercury went to a right-angle Holley carb; these are not too suitable for souping work, so we won't discuss them further.

Now our problem in carbureting a souped V8 is merely one of selecting venturi and jet sizes that will give us the desired mixture ratio and air flow under the various conditions with dual carbs, hot cams, porting, larger displacements, etc. However, since the carb is designed to meter fuel according to the flow rate through the barrel, adding carbs, cams, displacement, etc. doesn't upset this balance, and the carb just meters more or less fuel as the air flow through the barrel is increased or decreased. So you can just as well run standard jets and venturis as not.

However, since performance is more important to us now than gas mileage and idling, we might do well to juggle things a bit. Obviously our peak HP is a maximum when pressure loss through the venturi is low, and this means the largest possible venturi; on the other hand, for flexibility and acceleration at speeds below 2500 rpm, a smaller venturi is better. As a result, the old Stromberg 48 is best for high-output condi-

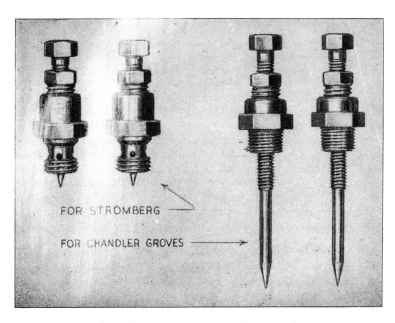

Fig. 7-3. Adjustable main jet sets for Ford-Merc carbs, pre-1949.

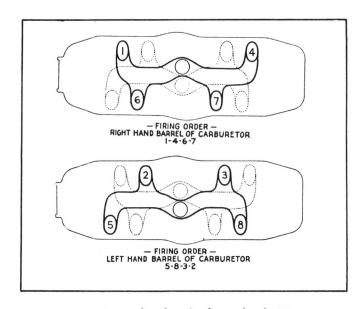

Fig. 7-4. Stock carburetion layout for the V8.

tions; for best performance around town or in drag racing, the Stromberg 97 or a late Ford model is better. For all-around use, however, the Strombergs are much more widely used than the later Chandler-Groves, as they are easier to work with and there appears to be a hair less air flow pressure loss.

Now a word about jet sizes. A number of optional jets are available at supply houses for all the Ford carbs ranging from 0.039 to 0.058 in. (and if larger sizes are desired, you can drill these out). For road engines, where cooling is not critical and gas mileage is a factor, a standard jet size around 0.050 in. gives good results. If you're running at high altitude, knock off 0.001 in. for every 2,000 ft. If you are after maximum fuel economy regardless of performance, try something in the low "40's" but keep an eye on the temp gauge. For highly-souped large-displacement road or track engines, the boys often use jet sizes in the high "50's" or low "60's" (by drilling out) for maximum possible HP and cooling.

Fuels other than gasoline will require other jet sizes under otherwise similar conditions, and you'll need to modify according to your fuel mixture proportions. Below is a list of our fuels with their required jet size relative to gasoline, rated "100:"

Gasoline - 100   Methanol - 195
Benzol - 90   Acetone - 142
Ethanol - 151   Ether - 157

If you're converting to methanol, as is the usual practice for sprint racing, jet sizes will run around 0.090 in., and we strongly recommend that you use one of the several conversion kits available at about $4.50, rather than try to do the whole job yourself, as quite extensive modifications are required for the large fuel flow.

Now a word about adjusting multiple carbs. In the first place, you should obviously use identical carbs and jets clear through, and your throttle linkage should be aligned to give exactly equal butterfly opening on all carbs (follow your kit instructions on this). Your idle adjustment is quite simple, use a manifold pressure hookup and adjust the idle valves so that the maximum vacuum is obtained (and this should be about 20″ Hg.). If you have a radical cam in it that won't idle, run it at low RPM and use a rubber tube as a stethoscope against the carb barrels, adjusting till you get the same sucking sound from each carb.

For checking your jets, about the only way you can do it is to accelerate your car at full throttle up to maybe 80 mph, declutch, shut the engine off, and then check the appearance of your spark plugs (assuming they're of the proper heat range). Use the following table:

TOO RICH	Base of plug sooty or wet
CORRECT	Base of plug slightly sooty; not too wet; light brown porcelain
TOO LEAN	Base of plug ash grey; burned, glazed brown porcelain

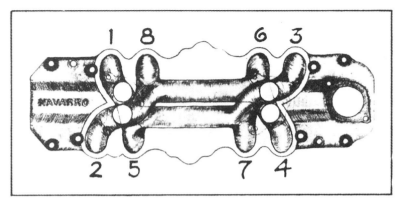

Fig. 7-5. Typical distribution layout of a dual manifold. The numbers at each port indicate the order of firing; notice that alternate suction strokes draw from opposite ends of separate manifolds.

Fig. 7-6. Cutaway view of a typical dual manifold showing the downflow into the ports.

If you follow our previous recommendations as to jet size, you shouldn't run into any trouble here.

MANIFOLDS

A very wide selection of multiple intake manifolds are available for the Ford-Merc V8, for every purpose, and ranging in price from about $30 to $45. They are laid out to equalize as much as possible the distance from the carb barrel to each port and to give even drawing impulses on

Fig. 7-7. Typical "Regular" dual manifold, allowing stock placement of generator and fuel pump.

Fig. 7-8. Typical "Super" dual manifold with special generator and fuel pump placement. This layout will give about 4% more peak HP than the regular.

each barrel, as determined by the engine firing order. (See Fig. 7-5). Two-carb manifolds are by far the most popular. These come in two types: (1) The "Regular" type, with the carbs coupled close at the center with high "risers," which allows stock placement of both the generator and fuel pump, and (2) the "Super" type, having the carbs set wide apart over the two sets of ports, with the generator bracketed over on one head and the fuel pump set at an angle, operated by a special pushrod.

Figs. 7-7 and 8 show the two types. As we learned earlier, one way to cut induction pressure losses is to shorten the length of the flow path; the Super manifold has much shorter flow paths than the regular and, as a result, has a peak HP margin of about 4%. For this reason, and since equipment is furnished with the kit for replacing the generator and fuel pump (at little extra cost), we can see no earthly reason for using the regular for any purpose. Incidentally, beware of the inexpensive close-coupled manifolds with LOW risers; in order to have the carb bowl clear the generator, the carbs are shoved way back, which gives unequal flow path length and distribution between cylinders. We have seen these installations in operation and would not recommend them.

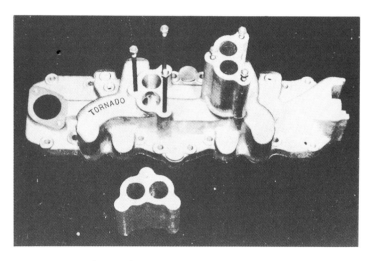

Fig. 7-9. Regular dual manifold with replaceable high risers; we get a bit more power with the low risers, but must bracket the generator on the head.

The Super manifold can be furnished with or without (or with detachable) 'heat risers" for evaporating the fuel mixture by exhaust heat, using the stock connections. Manifold heating expands the incoming gas, decreases its density, and robs you of maybe 5% of your HP. But by vaporizing part of the charge, we improve mixture distribution between cylinders and give a more flexible, smooth-running engine. We recommend heat-risers for the "conservative" and "medium" categories, but not for the hot road engine or racing.

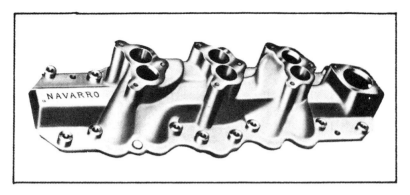

Fig. 7-10. Navarro triple racing manifold; these are designed for very large displacements or for running alcohol fuels.

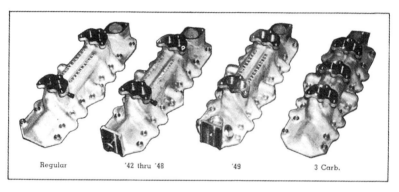

Fig. 7-11. Line-up of Offenhauser racing manifolds.

In general, a dual manifold is a pretty fair method of souping from the dollar standpoint, giving you 10-12% HP boost on the V8 for around $50 total investment. We recommend multiple carburetion for all engines.

But here is an essential point to remember: Since induction pressure losses decrease about as the square of RPM, your dual carbs will give you very little performance boost below 2500 rpm (see Fig. 7-12). If you're after that low-speed punch, hot heads and increased piston displacement are the answer — manifolds won't help much.

Now how about the three and four-carb manifolds? These are designed primarily for competition with alcohol fuels where very high flow rates are involved. With stock displacement, a triple manifold will raise HP only about 3% over a dual, and we don't consider it worth the expense and complication on a road engine. For hot V8 road engines over 280 cu. in. displacement, you might consider a triple, but for the others, just a Super

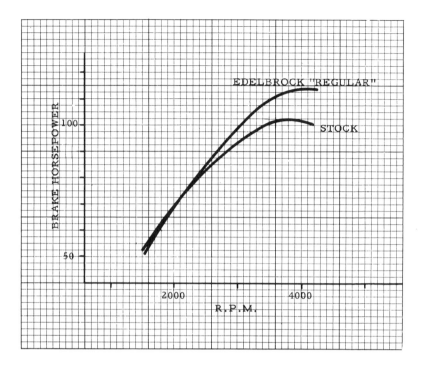

Fig. 7-12. Edelbrock curves showing HP increase given by a regular dual manifold (engines running without accessories). Notice no performance boost at low speed.

dual with or without heat-risers will work okay. For large competition engines, we recommend a quad.

Now a word about "porting" your manifold. Special compound manifolds are made to match up with the stock block ports. Obviously there is little point in porting the block if your manifold just overlaps the port -- in other words, manifold and block should be ported equally, with the port relief being tapered up into the manifold tube. This will also require your cutting a new manifold gasket. It's a good idea to consult your manifold manufacturer as to how much you can safely port his manifold casting.

PORTING AND RELIEVING

They tell us Leadfoot Louie's first real brush with "porting and relieving" happened way back when he was young and foolish. He poured two pints of Virginia Dare Port in the gas tank of his much-used '36 Ford, hopped in, wound it up to 6000 rpm in low gear, and relieved it of two rods and the right side of the crankcase! Then when Louie finally did find out what he was after, he ruined at least two blocks by going 3/8

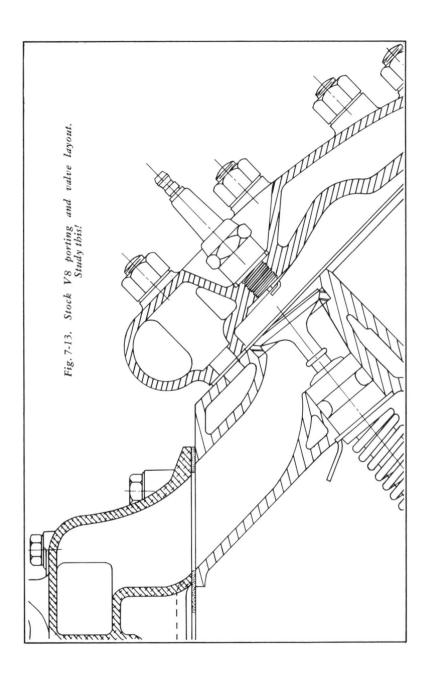

Fig. 7-13. Stock V8 porting and valve layout. Study this!

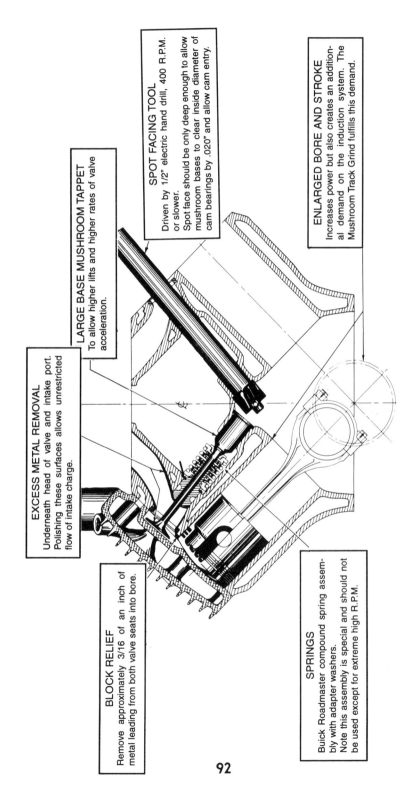

Fig. 7-14. Iskenderian drawing illustrating porting, relieving, etc.

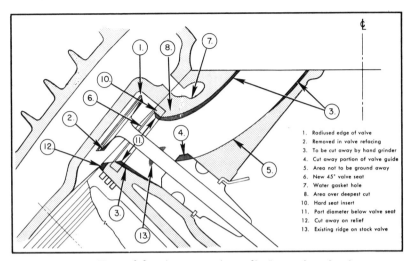

Fig. 7-15. Howard drawing on porting, relieving, and stock valve modifications.

1. Radiused edge of valve
2. Removed in valve refacing
3. To be cut away by hand grinder
4. Cut away portion of valve guide
5. Area not to be ground away
6. New 45° valve seat
7. Water gasket hole
8. Area over deepest cut
10. Hard seat insert
11. Port diameter below valve seat
12. Cut away on relief
13. Existing ridge on stock valve

in. deep on the port relief and ripping right into the water jacket. (Where does this guy get his block money?) But let's forget Louie and concentrate on doing the job right.

As you all know, "porting" refers to enlarging and smoothing the block ports — and "relieving" means to grind off the block face between the valves and cylinder to increase the gas flow path area through here (as high-compression heads always form a severe restriction at this point). Figs. 7-14 and 15 illustrate these steps.

Taken together, porting and relieving will boost peak HP about 5-7% on the V8, which is only fair from a dollar standpoint. However, when you get into large bores and strokes with the hotter engines, a good port and relieve job is essential to make full use of the larger displacement, since the stock induction layout throttles off part of the greatly increased gas flow. Porting and relieving are jobs you can do yourself if you've got a hand grinder, but many speed shops will do the whole job for about $35 — and do it right. So unless you're very experienced in machine work, we recommend that you have it done.

However, if you intend to do it yourself, here are a few pointers: Start by opening out the oval block ports about 1/16 in. around the edges and follow a path into the port about as shown in Fig. 7-15. There are many theories and methods for porting, so don't quote us on anything (notice that Figs. 7-14 and 15 don't agree on it!). Some fellows don't try to take off a lot of metal in their porting, but just smooth out the channel.

Use a fairly high-speed motor (around 5000 rpm) and a conical or roundnose grinding stone. The diameter of the stone should be about half the diameter of the port, and when grinding, rotate the stone around in the port in the direction opposite to the grinder shaft rotation. Be sure to taper in as you go down to avoid grinding into the water jacket — and be certain to get all your ports the same size and shape.

As for relieving, Fig. 7-16 shows the approximate contour to follow. The relief should taper down from the valves to a depth of 3/16 in. at the edge of the cylinder wall; don't go deeper than this or your top piston ring may ride onto the relief at high RPM and break it. Remember — in all your porting and relieving work — SMOOTH, EASY CONTOURS ALL THE WAY! (But we still suggest you shell out a few bucks and have it done.)

Fig. 7-16. Approximate contour to follow when relieving your block (light area); the relief edge will virtually follow the gasket line.

VALVES

Here is a field that has probably had less attention from V8 hot-rodders than any other. For one thing, the late V8 block, with its valve head diameter of 1.51 in., doesn't leave much room for larger valves, and the boys just figured that the slight gas flow improvement wouldn't be worth the trouble of fitting larger valves. We have no dynamometer figures at hand to answer that one way or the other.

However, by absorbing all the combustion chamber space with a larger intake valve, leaving the exhaust alone, some real possibilities appear. Several companies supply special oversize Ford valves, or you can adapt certain other stock valves with slight modifications. For example, the Ford 6 intake valve has a head diameter of 1.65 in. and can be used by shortening and modifying the stem. Or the DeSoto-Chrysler 6 valve, which has a 1.72-in. head, can be used with adjustable tappets, since total lengths are within 0.040 in. Also you can use an Olds "Rocket" intake valve (1.75 in. head) with adjustable tappets and some stem modifications to take the spring retainer; another possibility is to make up special retainers to fit the stock keepers to maintain proper spring length.

In the cases of fitting very large valves, the seat insert is removed on the Merc block and the port opened out below it; on the Ford, they just bore out the port approximately on the O.D. of the Merc insert. There are a thousand ways of doing the job, so we're not going to set any hard and fast rules. However, here's one thing to remember: Fit the valve with a narrow seat not over 1/16 in. wide; this not only increases your port area for a given valve diameter, but it gives a better seal. Also remember that you may have to fly-cut into the head for clearance when fitting oversize valves.

Now there are also some ways you can improve your stock valves. Fig. 7-15 shows the procedure. In the specific job shown, the valve refacing machine is set to 18° and the underside of the head is ground off till the valve seat is about 0.025 in. wide (0.040 in. for exhaust). Next the port diameter is ground out 0.105 in. (to 1.445 in.) with a hard seat stone dressed to 72°; then, with a 2-in. O.D. stone dressed to 17°, the

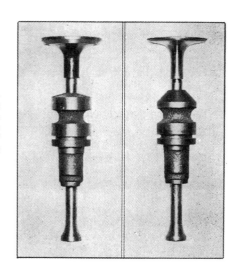

Fig. 7-17. Commercially reworked stock exhaust valves and guides (stock contour is on the left). The underside of the valve is shaved and the part of the guide extending into the port is tapered.

tops of the valve seats are refaced to the proper seat width. This is just a suggested procedure — and it is said to raise HP about 3% at the peak. At least one company sells stock Ford valves and guides modified in this way. (see Fig. 7-17). Incidentally, if you are reworking stock valves, use exhaust valves clear through as they are made of tougher material.

From a dollar standpoint, increasing valve and port size is not such a hot way of boosting performance, so here are our valve recommendations: For the "conservative" and "medium" categories, the stock layout will be sufficient. For the hot road and track engine, we suggest some changes, either modifying the stock valves or fitting oversizes. For super-competition jobs, overhead valves are in order, as mentioned earlier.

SPRINGS AND TAPPETS

Here are some seemingly insignificant parts that don't pay off in HP — but we must have them right or the whole souping job goes sour. Our valve springs are a critical problem with reground camshafts because of the greatly increased opening and closing rates (acceleration), and because of the higher RPM involved. Just as the case with pistons and rods, the weight of our valves, springs, retainers, etc. sets up considerable "inertia" forces when they move up and down; when these accelerations are increased by different cam contours and RPM, the tendency is for the stock springs to begin to "float" (fail to follow the cam on the closing side) at speeds as low as 4000 rpm.

Stock Ford-Merc springs have a tension of 82 lbs. with valve open, and it is generally agreed that this should be increased 50% for any reground cam to avoid floating up to 6000 rpm. We have several alternatives. We can use stock Lincoln-Zephyr springs without any changes; these have an "open" tension of 117 lbs. Or for very high RPM, you can use stock Buick 70 dual springs by using adapter washers (see Fig. 7-14); the rated tension here is 170 lbs. Probably the best bet is to invest $4 in a set of special racing springs for the V8, available from several supply houses. At any rate, you should have at least 110 lbs. open spring pressure for any reground cam (but remember, the more pressure, the more friction and wear on the cam and tappet). And incidentally, when fitting the springs, if the coils are closed at one end, that end should butt against the guide — not the retainer!

How about tappets? Unfortunately Ford never used an adjustable tappet, which means we must use some method of taking up the amount ground off the cam heel when we use a hot cam. You can build up the end of the valve stem by welding or you can shim it up. However, the recommended procedure is to sink about $14 in a set of special adjustable tappets (see Fig. 7-18). And here's a tip on that: To ease valve clearance adjustment, drill a small hole through the tappet boss near the base; you can insert a pin or rod through this hole to catch the slot in the tappet, so it can be held solid while you adjust (otherwise you have to lock the tappet separately with a special tool each way you turn the nut).

Fig. 7-18. Witteman adjustable tappets for the V8, with self-locking adjustment.

THE CAMSHAFT

Of all the steps we can take to soup our stock engine, nothing will pay off in HP per dollar like messing with the camshaft — and nothing will kill our performance at low RPM quicker! When Leadfoot Louie first heard about cam grinding, he ripped his cam out and went to work on it with a file. That didn't work, so he took a hand grinder to it. Somehow he got the engine started again, but you should have heard it. It's ridiculous, but idling at 1000 rpm, it sounded like all 8 cylinders were running a race to see which could fire faster — and the leader stumbled! We know of guys who have ground their own cams and done a good job, but there are too many Leadfoot Louies around for us to encourage it.

Now just what is our problem here? Basically, of course, the idea is to get the maximum possible weight of fuel mixture into the cylinder on the intake stroke, at all RPM. We talked a little about valve timing back in Chapter 3, and there we learned that our timing must necessarily be a compromise. We can't raise HP at both high and low RPM through valve timing.

The reason is obvious: All we do when we fool with cam timing is to try to take advantage of the inertia of the gas flow — and inertia

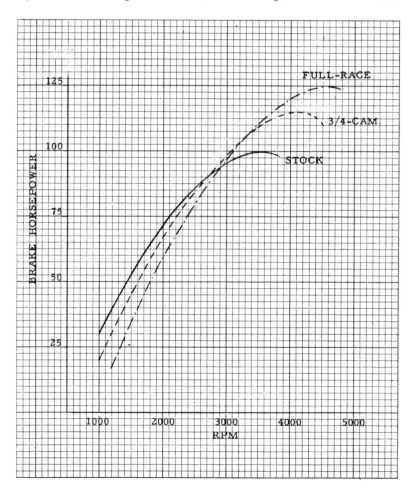

Fig. 7-19. Graph showing the effect of different cam timings on performance at all RPM; no other changes.

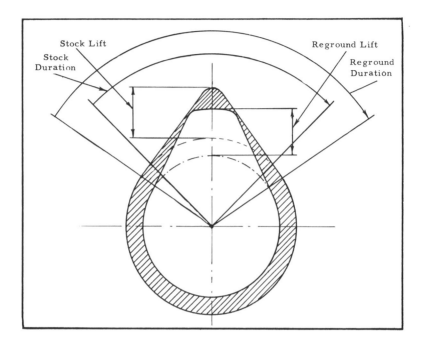

Fig. 7-20. Changing the cam contour by regrinding. By grinding off the shaded portion, valve timing and lift can be altered in most any way we wish.

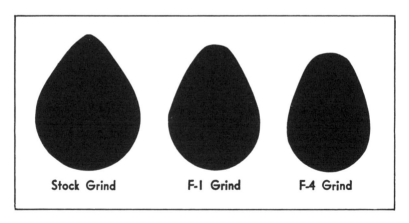

Fig. 7-21. Actual photographs showing cam contour with two Weber regrinds (the F-4 is a track grind).

decreases as the square of the flow rate (RPM). Or to put it simpler, to get a good volumetric efficiency at high speeds, we must leave the valve open a longer period to take advantage of the inertia or momentum of the incoming gas to further fill the cylinder; but at low RPM, there will be practically no inertia and the late closing will just cause some of the gas to be pumped out of the cylinder, and compression will be delayed till the cylinder is sealed.

So we see that, what helps us at high speed hurts us at low speed. Stock timing, then, has to be a careful compromise that will give us peak torque at around 1500-2000 rpm, and maybe 30 hp at 1000 rpm. This means very modest volumetric efficiencies of around 60% at 4000 rpm.

Now when we seek to modify that same stock cam in our souping work — WE HAVE NO CHOICE BUT TO BREAK THE COMPROMISE. In other words, by boosting peak HP through cam timing, we automatically cut off a slug of horses at 1000 rpm. Fig. 7-19 shows this very clearly. It's essential that we get this firmly in mind. Some manufacturers claim improved performance at both ends of the speed range — but that's a debatable question. It can't be done with a simple regrind job.

For a track engine, low-RPM output is of no importance anyway, but on a road engine, safety in traffic flow demands that we have at least a little something at the bottom end. Remember this. And here is another point: A given cam grind (like a 3/4 grind) will give better low-RPM performance as the bore and stroke of the engine are increased, since the cam timing can be better utilized in handling the greater gas flow. That's because the various grinds are rated on a basis of stock displacement. In other words, a full-race grind might idle pretty well on a 286-cu.in. block, but would idle mighty rough or not at all on a stock block (239 cu. in.). Keep this in mind when planning cams.

Now just what do we do to our stock camshaft to "hot it up," and how far can we go? Obviously what we want is longer valve open periods, greater lift, and quicker opening and closing. This will give us the maximum valve discharge area for the maximum time period — and will give us the maximum gas induction on the intake stroke at high RPM. We can get all these things by simply grinding down the stock cam. Fig. 7-20 shows what is done. The price for the job averages about $30 for a V8 camshaft, and there are hundreds of shops around the country that are equipped to do it.

Now obviously we will need a wide variety of different 'grinds" to fill the many souping requirements. Certainly John Q. Public can't use the same cam for a Sunday afternoon drive that Joe Blow uses at Squeedunk Speedway. So the grinding companies have more or less agreed on a standard nomenclature for the various grind categories. Though individual grinders vary somewhat in their exact timings, here is a run-down of our different grinds:

SEMI-RACE GRIND: This is the mildest grind, timed about 20-60 intake and 55-15 exhaust (the degrees before and after dead center). It's quite an improvement over stock, however, gives a peak HP boost of

FORD V8 VALVE TIMINGS

(While reground cams vary somewhat with each supplier, common usage classifies the various types (approximately) in accordance with the following table for the Ford V8)

	LATE STOCK	EARLY STOCK	1949 MERC.	1/2 RACE	3/4 RACE	FULL RACE	SUPER (WINFIELD)	SUPER (HARMAN-COLLINS)	SUPER 'H' (HARMAN-COLLINS)
INTAKE OPENS (Deg. before T.D.C.)	0°	9½°	10°	21°	23°	26°	24°	28°	30°
INTAKE CLOSES (Deg. after B.D.C.)	44°	54½°	50°	59°	62°	64°	68°	67°	78°
EXHAUST OPENS (Deg. before B.D.C.)	48°	57½°	50°	54°	56°	59°	68°	61°	64°
EXHAUST CLOSES (Deg. after T.D.C.)	6°	6½°	10°	16°	19°	21°	24°	24°	26°
INTAKE DURATION (Deg. crankshaft)	224°	244°	240°	260°	265°	270°	272°	275°	288°
EXHAUST DURATION (Deg. crankshaft)	234°	244°	240°	250°	255°	260°	272°	265°	270°
OVERLAP (Deg. intake and exhaust are open at the same time)	6°	16°	20°	37°	42°	47°	48°	52°	56°

Fig. 7-22. Typical cam timings with reground cams.

CAMSHAFT TIMINGS

WINFIELD	Intake opens BTC	Intake closes ABC	Exhaust opens BBC	Exhaust closes ATC	Lift	Clearance Intake	Clearance Exhaust
WINFIELD							
60' Super	24	68	68	24	.250	.012	.014
60' R-11-S	16	60	56	16 In.	.268	.012	.014
				Ex.	.256		
Super No. 1-A	28	72	62	22 In.	.350	.012	.014
				Ex.	.330		
Super No. 1	24	68	66	22 In.	.325	.012	.014
				Ex.	.315		
Super No. 1-R	22	66	62	18 In.	.315	.012	.014
					.305		
Full Race	18	62	62	18	.305	.012	.014
Three Quarter	18	62	58	14 In.	.305	.012	.014
				Ex.	.295		
Semi	14	58	58	14	.295	.012	.014
HARMAN - COLLINS							
60' Super	22	68	70	20	.256	.010	.012
60' Midget	24	58	68	16	.256	.009	.010
Super "H"	30	78	60	28	.350	.018	.015
Super	24	68	66	22	.320	.011	.015
Full-Race	19	59	59	19	.320	.011	.015
Three Quarter	19	59	54	14	.320	.011	.013
Semi	16	54	54	16	.320	.011	.013
WEBER							
60' Boat S-2	23	78	72	30	.265	.010	.010
60' Midget S-1	22	70	72	20	.265	.010	.010
"R" Grind	25	60	60	25	.290	.012	.012
F-1 "	22	62	66	18	.312	.012	.013
F-2 "	30	65	70	30	.320	.013	.014
F-3 "	42	78	84	38	.320	.013	.014
F-4 "	32	71	74	28	.340	.014	.016
F-4-A Grind	32	71	74	28	.330	.014	.016
F-4-B "	32	71	74	28	.315	.014	.016
Track Grind	18	62	64	16	.330	.013	.015
SMITH & JONES							
60' Boat	22½	63½	63½	22½	.254	.012	.013
60' Midget	15	59	58	14	.254	.012	.013
85—No. 252	16	56	56	16	.312	.012	.013
85—No. 260	21	59	61	19	.312	.012	.013
85—No. 266	24	62	62	24	.312	.012	.013
85—No. 272	27	65	62	24	.312	.012	.013

Fig. 7-23. Some camshaft timings of different grinding companies.

about 10%, and yet retains good low-RPM performance. This grind is okay for Aunt Effy, but one could get something stronger for the same price, such as:

THREE-QUARTER RACE GRIND: This is the next in line, timed about 23-62 intake and 56-18 exhaust. Here is the best one for road engines with stock displacement; it gives a power increase of about 15% and still has fairly good lugging ability at low speeds (though you understand that none of these grinds will outpull the stock cam below about 2000 rpm)

FULL-RACE GRIND: This grind is in "no man's land" — too much for John Q. and not enough for the track man (maybe Leadfoot Louie could use it!). It is timed about 26-64 intake and 59-21 exhaust, and gives a HP boost of near 20% at the peak. It doesn't have much below 1000 rpm, but it's pretty effective in a hot road engine of 250-270 cu. in.

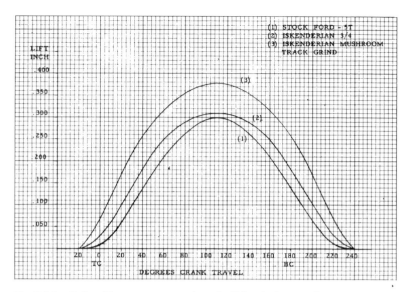

Fig. 7-24. *Valve lift curves comparing the Iskenderian mushroom cam with a ¾ and stock cam. This deal pays off in some 5% HP margin at the top.*

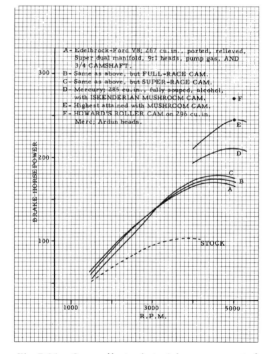

Fig. 7-25. *Some effects of special cams on souped engine performance.*

SUPER-RACE GRIND: This is a track grind timed about 28-70 intake and 61-25 exhaust. It has all its effect at high speed but is of little value below 1500 rpm with stock displacement; however it might be used successfully in out-and-out hot rod of over 270 cu. in.

SUPER-SUPER RACE GRIND: Here is a puzzler. As soon as we list the timing, somebody will come up with a "super-super-super" grind and make us obsolete in a cloud of oil smoke and flying pistons! Anyway, this most radical of grinds today might be timed 30-80 intake and 65-25 exhaust. It is designed strictly for track racing with very large displacements, and would give a HP boost of probably 25-30% if used with stock bore and stroke. The torque curve on this one starts dipping below anywhere from 3000 to 4000 rpm, so don't try to use it on the road.

Some accompanying charts show typical valve timings with reground cams. You'll notice also that valve lifts are not too much greater than stock because of the low combustion chambers on the high-compression heads; they average about 0.315 in. Tappet clearances to be used with these cams will vary somewhat according to the grinder's recommendations, but average 0.012 in. on the intake and 0.014 in. exhaust.

Incidentally, here's a tip: You can "cool" your cam considerably by just increasing your tappet clearance; if you have stuck in too hot a grind and can't get along with it in town, try boosting your tappet adjustment a couple thousandths (it's noisier, and HP suffers a bit, but you have no other alternative).

Now a word about those super-special cams — the "mushroom" and roller-tappet setups. These are based on the principle that the height of the cam "toe" must be low enough so that it doesn't fall outside the tappet base as it comes around — which, of course, limits both the amount of lift you can get and the total valve open periods. In the Iskenderian mushroom setup, the stock 1-in. Ford tappet is replaced with a conversion unit (see Fig. 7-14) and a special cam is used. Fig. 7-24 shows comparative valve lift curves and Fig. 7-25 shows the remarkable HP dividends achieved with this layout.

In the Howard roller-tappet setup, a special tappet is used with a roller cam follower (which is the equivalent of an infinite tappet base width), and the cam contour can be given any combination of lift, duration, and valve rate that you want! For example, you can get 0.500 in. lift and 300° intake duration, compared with about 0.320 in. 280° for conventional track grinds. So far, these special mushroom and roller cams appear to have a HP margin of about 5%, but reliable dynamometer data is scarce, which makes it difficult to evaluate them (for example, Howard's 270 hp on the Ardun-Ford was achieved with some nitro-methane in the fuel). Anyway, the new cams are expensive, ranging nearly $100 for the setup, but for serious competition, we suggest you investigate them carefully.

Now here are our cam recommendations: Forget the semi grind in all cases. For a "medium" road engine with stock bore and stroke, the 3/4 is best. If you're going pretty hot with your road engine, try the full-race cam if your displacement is below 270 cu. in., and a super-race above that. This idea may sound a little rash for the road, but it will definitely work

— though, of course, low-RPM performance will no more than get you around. For competition engines, your cam selection will depend on the type of racing you're doing. For sustained high-RPM work, the sky's the limit. If operation will be over a wide RPM band, such as road racing, the milder grinds will be better (Fig. 7-25 shows some power curves of different grinds on the same souped engine). So use your head when you select a cam — but don't be too careful!

THE EXHAUST SYSTEM

Actually, your exhaust system will not be nearly so critical as to gas flow refinements as the induction side, because your flow here is pumped under a pressure of maybe 80 lbs. (when the exhaust valve opens), whereas induction is under only $14\frac{1}{2}$ lbs. atmospheric pressure. However, it's still true that you can raise peak HP some 10% over the stock setup if you take care of a few details.

The exhaust layout has always been a problem on the V8, and there's quite a bit we can do to help things. We have already discussed exhaust valve modifications, so let's consider the exhaust passages in the block. The two center cylinders on each bank feed exhaust gas directly at each other into a common channel between the two cylinder walls. With very high gas flow rates, this "collision" of the flow from both cylinders naturally sets up some back-pressure.

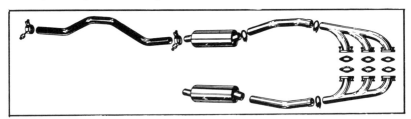

Fig. 7-26. Complete Belond dual exhaust system for the Ford-Merc with racing-type headers. This setup practically eliminates 'back-pressure" and runs about $50.

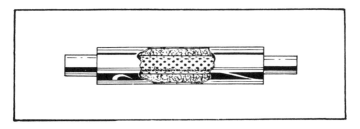

Fig. 7-27. Cutaway showing the construction of a "Hollywood" muffler. The straight-through pipe is perforated and packed in steel wool; it deadens rather than muffles the exhaust note.

There are now some cast iron baffles on the market that you can slip up through the exhaust passage and fasten in place with a head stud, that will separate this flow. These are a good investment for a highly-souped engine, since it also helps the cooling problem somewhat. Some fellows also open out the exhaust heat riser holes in the block and install auxiliary tubes connecting these holes with the main exhaust pipes.

And then there is the problem of the external exhaust system. The stock layout, with both banks feeding into a single pipe, muffler, and tailpipe, offers a great deal of resistance at high flow rates, and robs about 7% of our HP — besides greatly aggravating the cooling problem with high-compression heads.

Fortunately there's a wide selection of open dual exhaust systems for the V8 on the market that will satisfy the Law as to muffling, and won't take more than 1% of the developed HP (and none under most operating conditions). The racing-type headers run about $30 a pair, and the two mufflers and tailpipes cost around $20 (see Fig. 7-26). These mufflers are of the "straight-through" type, with a perforated tube imbedded in a steel-wool packing; they offer practically no flow resistance, since they only deaden the exhaust sound, rather than muffle it. You have seen the ads — a gal waving at a guy driving by with his twin pipes saying, (quote) "m-m-m-m-m-m" (unquote)! She would really get a kick out of Louie — half the time he tears around with his block ports running open! Anyway, we'd say that some sort of dual exhaust system, whether using stock mufflers or with or without headers, is just about a MUST for any road engine above the "conservative" category. In fact, that $50 investment is a good way to start your souping and add 10% to your HP without touching the engine. (Don't forget, however, that this doesn't do too much for your output at low RPM).

So that about concludes this chapter on the all-important induction and exhaust systems. Here we are with a nice round, firm, fully-packed fuel charge in our cylinder. The next job is to fire it.

Having made an official run of 160 mph at the 1951 S.C.T.A. Bonneville Trials, this car has 239 cu. in. Merc engine equipped with Roots-type supercharger, is an outstanding drag machine.

A unique installation of a reworked GMC supercharger with chain drive on a 274 cu. in. Mercury plant. (Road Runners).

CHAPTER 8

IGNITION

LEADFOOT Louie's got a terrific idea for ignition that will solve all the souper's problems! He describes it thusly: "I'm tossing out the spark idea and building up a special plug with a filament that will glow red-hot on a current of 5 amps. When the temperature on the compression stroke reaches a certain point ... poof ... automatically. No troubles with spark lag and cutting out. There's one problem, though — I'll have to run double generators to pump the 40 amps needed!"

Who are we to say Louie's crazy? But we definitely know another guy who's barking up the wrong tree. He was trying to run a full-race V8 mill at high RPM in a 225 hydroplane; he had $9\frac{1}{2}:1$ heads, triple carbs, super-race cam, a full house — except STOCK IGNITION. And he was wondering why the thing was loggy and cut out at 3500 rpm!!! (He was sure it must be carburetion.)

This problem of ignition is a vital part of our overall souping job. All the heads, cams, and carbs in the world aren't going to put torque to the flywheel if we don't fire that charge in every cylinder at exactly the right instant. Because ignition doesn't pay off in a flock of extra horses and because fabulous claims aren't made for the special equipment, the inexperienced rodder is apt to forget it in his plans. It's "souping suicide" — don't do it!

BASIC FUNDAMENTALS

Before we can intelligently approach the problems involved in ignitioning our souped engine, we must first understand the basic requirements and the limitations of the original stock layout. And it will help if we pause right here and get one thing firmly in mind: Electric current acts very much like a liquid or gas flowing in a closed system. In other words, we have to work with concepts like pressure, friction, momentum, etc. — only with electric current we use terms like volts, amperes, line drop, and impedance. But the idea is the same. When electric current flows through a wire, it requires a pressure or voltage to move it; the "friction" of the flow appears as a voltage or pressure drop; and electric current has a property something akin to weight, so it possesses a definite inertia and you can't start it or stop it instantly! Keep these facts in mind when you study ignition.

Now we're not going into a lot of detail here on ignition fundamentals because we're sure you all know how a battery-coil system works. Let's just look at our basic problem — which is, of course, to get a good hot spark, with no lag, under all conditions and at all RPM.

Now in the first place, we've got two strikes on us from the start because, in all our usual souping work, we're actually plotting against the ignition system! Combustion theory teaches that the current voltage re-

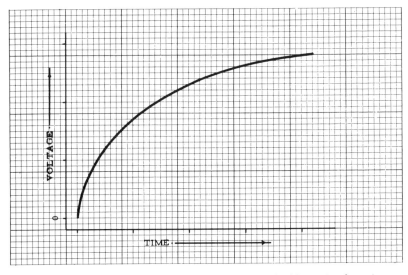

Fig. 8-1. Generalized curve showing how voltage builds up in the primary coil when the breaker points are closed.

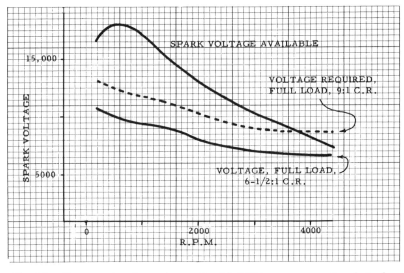

Fig. 8-2. How spark voltage drops off with RPM on a conventional single-breaker ignition, because of less time available for saturating the primary coil; the lower curves show voltage required to throw a spark.

quired to jump a given spark gap is proportional to the gas density between the electrodes. Therefore when we port, relieve, install hot cams and raise compression ratio, we just boost the compression pressure and increase the voltage required to throw a spark.

To be more specific, assuming constant manifold temperature, our required spark voltage theoretically increases directly with volumetric efficiency and compression ratio. This is rougher than it sounds. For example, suppose, by careful souping, we raise volumetric efficiency at a given RPM by say 20% and we raise compression from 6½ up to 9:1. Then our required spark voltage is increased: $1:20 \times 9.0/6.5 = 66\%$. It isn't quite this bad in actual practice, but it's still a big boost to ask our stock ignition to handle.

And this is only half the problem. There is a basic limitation of the battery-coil ignition system that's working against us too — and that is the fact that the available spark voltage drops off fast as the RPM goes up. Here's why: As we mentioned earlier, electric current has "inertia," and we can't just close a circuit and have it start and reach full flow instantaneously. Therefore when the breaker points close and start the current from the battery to the coil, the voltage builds up gradually in the primary winding — something like Fig. 8-1. With a constant feed pressure of 6 volts from the battery, the time required to build up primary voltage is virtually independent of the RPM.

So we find that, as speed increases, the points are closed a shorter and shorter time, and the primary current is being chopped off before it has time to get rolling. This, of course, reduces the output of the secondary coil to the spark plug in proportion. Fig. 8-2 shows how spark voltage drops with RPM for a typical stock car installation. You stick this same layout on a souped engine with 8 or 10:1 compression, wind it up over 4000 rpm, and what happens? The spark voltage is not sufficient to jump the gap and the engine sputters and cuts out. It's hard to say just where a certain stock ignition will fail at such and such a compression ratio with a certain fuel — but it's a cinch that something will have to be done with the stock setup for most souping jobs.

Now just what can we do? From the standpoint of reworking stock ignition, two major possibilities appear: (1) Use a special high-output coil that will give a peak spark pressure of around 30,000 volts, instead of the usual 20,000 for stock, and (2) use two separate breaker-coil systems so that we can double the time the breakers are closed at a given RPM.

The first possibility, the high-output coil, will help up to a point, but even that's not enough for a highly-souped engine. On the other hand, a complete answer has been found in the "dual" system. By using two coil circuits, we can put half the cylinders on one and half on the other, thereby doubling the time available for saturating the primary coil at a given RPM. This system has been found completely satisfactory for firing the 8 cylinders of the Ford engine on 6 volts battery pressure under all souping conditions.

Now let's get down to actual cases:

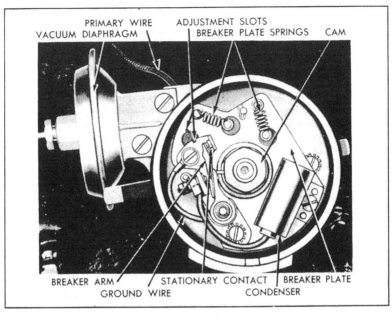

Fig. 8-3. Breaker assembly of the "single" ignition on the late post-1949 V-8's, using vacuum advance.

Fig. 8-4. High-output 30,000-volt D.S.M. coil. These will take care of the ignition on your milder souping jobs.

Fig. 8-5. Special Lucas (British) single-coil ignition for the V8. This uses a 4-lobe cam with two breakers and can be rotated around the mounting bolts for advance adjustment (automatic centrifugal advance built in).

REWORKING STOCK IGNITION

Before 1949, the Ford and Merc employed a very efficient little ignition system of the "double-breaker" type. This employs an 8-lobe cam, but there are two breakers connected in parallel, located in relation to the cam in such a way that their "closed" periods overlap; the total cam rotation angle from the time the first breaker closes till both are open is 37°— whereas with just one breaker on an 8-lobe cam, it would be about 27°. Here is a clever attempt to get a little longer charging time and still have enough point gap to prevent arcing. It's a beautiful layout for stock conditions (as witness Chrysler's adoption of it for their terrific 180-hp FIRE-POWER V8), but the cam angle still isn't nearly enough to fire a hot engine in the 5000 rpm range.

Early pre-war attempts to improve the V8 ignition included fitting heavier springs on the breaker arms and sometimes the use of lighter arms, to prevent floating at high RPM — which, of course, didn't solve the basic problem. Today there are on the market several simple kits for improving pre-1949 V8 ignitions through the use of special points, breaker arms, springs, etc. As mentioned, these things will fix it so you can wind your stocker up to 6000 rpm (since the old Ford ignition had a poor cam shape that invited floating), but these steps won't help any to cope with high compression, etc.

The same goes for the post-1949 ignitions. As you know, with the introduction of a new basic V8 engine in that year, the ignition system was changed entirely. It now uses an 8-lobe cam with only one breaker, and vacuum spark advance (see Fig. 8-3). This setup has a cam dwell of only 28° and is no good for our purposes (in fact, it will barely handle the stock job!). There are kits available for converting this to the old double-breaker layout, and complete exchanges can be had.

How about converting the stock ignition to a true dual coil layout? There are no kits on the market to enable you to do this yourself. We know of an experienced mechanic who holed up in his machine shop for five days, and finally came up with a workable unit. We don't advise you to try it unless you've got the brain of an Einstein and the patience of Job!

About the only possibility left for "beefing up" our stock ignition is to fit a more powerful coil. This is a simple matter and it will help a lot if you're not too deep in your souping. There are several good high-output (30,000-volt) coils on the market at around $12 — such as D.S.M., Mallory, Bosch "Big Brute," etc. These are a good investment for the mild souping jobs.

While on the subject of stock ignitions, a word about adjusting spark advance. As we mentioned in Chapter 6, when we work with high-compression cast iron heads on the V8, whether they be milled stock or special units, we are very apt to run into bad knock or "pinging" at compressions over about $7\frac{1}{2}:1$. In actual practice, things seem to be worse when accelerating hard at low RPM. Much of our trouble can be eliminated by merely retarding the spark.

All the pre-1949 double-breaker distributors had an automatic centrifu-

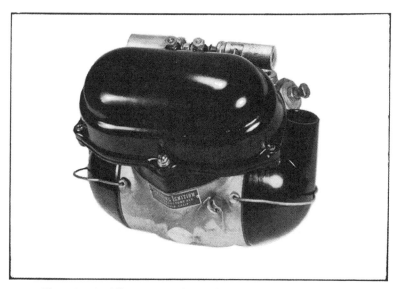

Fig. 8-6. *Spalding converted Lincoln-Zephyr dual ignition for the pre-1949 V8.*

Fig. 8-7. *Spalding converted 59A dual ignition for the post-1948 V8.*

gal spark advance mechanism that was retarded according to engine load up to about 1500 rpm by a vacuum brake; with this layout in stock adjustment, advance varies anywhere from 4° at low speeds to 26° at 3500 rpm. This works okay under stock conditions or with aluminum heads, but we'll probably need to modify with high-compression iron heads. The first thing to do is to experiment with tightening down the vacuum brake adjusting screw; this will hold things back a little at low speeds without cutting peak power. If you still get a bad knock after this, your only alternative is to rotate the breaker plate through the timing adjustment screw on the distributor body, and then lock it; in this way you can retard spark up to about 5° over the stock setting, but the retard is constant through the full RPM range and your peak HP suffers some.

On the late post-1949 ignitions, using a single breaker, there is no centrifugal advance, with all spark control being through manifold and carb vacuum; the stock advance range is from 2° to 18° at 4000 rpm. As mentioned before, this whole deal isn't much good from a souping standpoint and should be converted in one way or another. However, if you are bothered with knock (though this unit isn't as critical in this respect as the old ignition), you can rotate the breaker plate by loosening the clamp under the distributor body; there is no vacuum adjustment. Experiment until you find the point of maximum advance without excessive knock, since every degree of retard whacks off horses at the top end.

CONVERTED DUAL SYSTEMS

Here is an answer to every souping ignition requirement — at a reasonable cost. This "dual" idea was first tried out before the war in backyard shops, where it proved itself at once (though the old V8 breaker cam shape still tended to float). Since the war, a number of companies have brought out complete units ready to stick on your block. These are converted from stock Ford or Lincoln-Zephyr ignition units; some are designed to take the stock Lincoln double coil fixture, while some have universal coil mounts (one is special from the ground up).

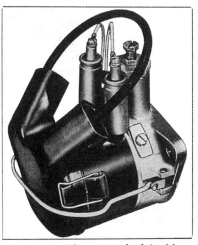

Fig. 8-8. Spalding 59A dual ignition.

Fig. 8-9. Kong dual ignition for the pre-1949 V8. This unit is "special" from the ground up and is one of the most rugged of the dual systems available.

Fig. 8-11. Roemer dual conversion of the pre-war Ford 21A ignition, for pre-1949 V8's.

Fig. 8-10. Speedsport "R-W" dual conversion of the post-1949 ignition. This setup uses the stock Ford points, condensers, coils, and vacuum advance, but has a 4-lobe cam, 70° cam dwell, and is guaranteed for 6000 rpm.

Big names in this field right now are Spalding, Kurten, Kong, and Roemer, plus a number of general speed equipment manufacturers who convert their own units in limited quantity. Prices vary generally from $45 to $60; the ignitions have a cam angle of 68° to 80°, are guaranteed to throw a spark up to 8000 rpm, and they're completely satisfactory for almost any souping conditions.

MAGNETOS

The magneto doesn't hold the place that it used to in the speed field. Time was when you couldn't do a thing with an engine without first sticking on a "mag" to handle the ignition problem. But today, with our highly-developed dual breaker-coil systems, a mag is only for the most extreme conditions.

As you know, a magneto is nothing more than a conventional breaker-coil ignition unit, except that the current is supplied by a built-in generator instead of a battery. It's all housed in one unit and can be conveniently driven off the stock distributor shaft. The major advantage of a mag, aside from some 30 lbs. of weight saved by eliminating the battery, is in its voltage output characteristics; since the armature cuts the magnetic

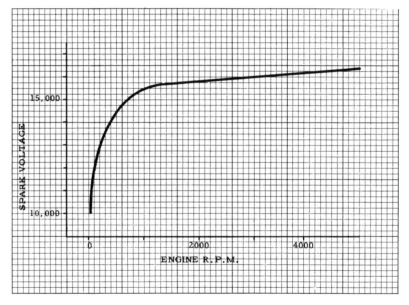

Fig. 8-12. Typical output curve for a Scintilla-Vertex magneto; electrical inertia effects level the curve off at high speed, but there is no voltage drop as with battery-coil ignition.

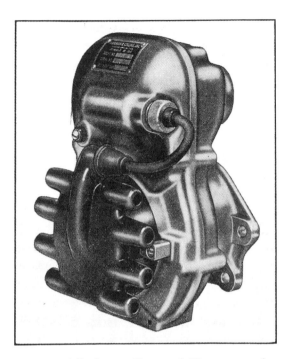

Fig. 8-13. The famous Harman-Collins magneto for the V8; it fastens right on the nose of the block in place of the stock distributor.

Fig. 8-14. A typical Scintilla-Vertex mag. These have automatic centrifugal advance and are available in road or track models.

field faster as the RPM increases, we have the delightful situation where our spark voltage actually increases with speed (though electrical inertia effects cause it to level off at high RPM's). Fig. 8-12 shows an output curve for the Scintilla-Vertex mag.

Actually, a magneto is the perfect answer to our souping ignition problem. But since the dual breaker-coil systems handle the job okay, and since a mag costs two or three times as much, we could only recommend one for the most severe competition conditions. Several companies, notably Barker, Harman-Collins, and Vertex, supply complete magnetos ready to bolt on the V8 block and go (see Fig. 8-13,14, and 15). These are all stock commercial mags with special adaptor equipment. Prices range from about $100 to $130.

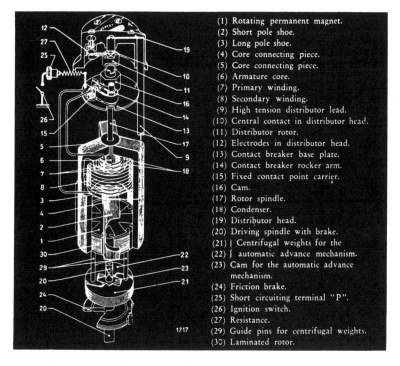

Fig. 8-15. *Diagram drawing of the Vertex mag showing main working parts.*

SPARK PLUGS

Any spark plug we can stick into the head can handle our firing — for maybe three minutes! But if we want a plug that will stay right in there blasting for a 20-lap feature, or 20,000 miles on the road, we're go-

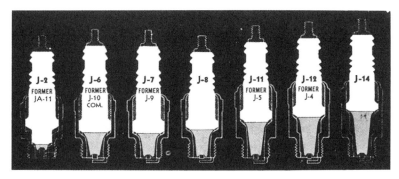

Fig. 8-16. Drawing illustrating spark plug heat range; "cold" on the left, graduating to "hot" on the right.

PROPER HEAT RANGE APPLICATION after appreciable operation, is shown below:

REGULAR FUELS

Rusty brown to grayish tan powdery deposit on firing end of plug; normal degree of electrode erosion.

LEADED FUELS

White powdery or yellowish glazed deposit on firing end of plug; normal degree of electrode erosion.

NOTE: Deposits characteristic of leaded fuels, sometimes called "encrustments", do not interfere materially with spark plug operation, and should merely be cleaned off at regular service intervals.

WORN-OUT PLUG

Caused by normal service beyond life of plug. Spark plugs should be replaced every 10,000 miles for maximum economy and performance.

PLUGS TOO HOT FOR THE TYPE OF SERVICE GENERALLY APPEAR AS FOLLOWS:

PLUGS TOO COLD for the type of service are shown below:

OIL FOULED

Wet, sludgy deposit on firing end of plug; negligible degree of electrode erosion.

GAS FOULED

Dry, fluffy deposit on firing end of plug; minor degree of electrode erosion.

White, burned or blistered insulator nose; badly burned and corroded electrodes.

Fig. 8-17. Photos showing the usual appearance of the plug tip under different operating conditions.

ing to have to be as careful about selecting it as we are with our heads, cams, and distributors.

Plugs are a critical factor in engine operation. In our souping work the problem becomes more acute because of the very high temperatures developed in the cylinder. With a stock engine, peak gas temperatures under full power might reach 3,000° F. during combustion, and the average cylinder head surface temperature could be 300°; when we soup that same engine and boost compression to say 10:1, our peak gas temp jumps to some 4,500°, and the surface temp in proportion!

Under a blast like this, a good stock plug will quickly heat up to a point where it will touch off the fuel mixture on the compression stroke long before the spark is timed to fire. This is called "pre-ignition," and it's deadly to engine parts as well as HP. Not only this, but an overheated plug will burn out in a hurry. So we see that spark plug heat range is a vital factor — and we can't necessarily depend on the stock plug type.

Now the "heat range" of a plug can be easily controlled by varying the length of the heat flow path from the electrode tip around through the body and gasket to the cylinder head. In other words, by shortening the distance through which the heat must pass, we decrease the flow resistance, increase the heat flow rate, and allow our plug tip to run cooler. Fig. 8-16 shows one type of Champion plug in seven different heat ranges. For the "cold" or "hard" plugs on the left, the flow path is very short, graduating to the hotter or "soft" plugs on the right.

Selecting the proper plug for our souped engine is simple, with such a broad range to choose from. Since 1938, the V8 has employed 14-mm plugs (18-mm before that), and the present Ford-Merc stock jobs use Champion H-10 types with 0.025 to 0.030 in. gap. This plug might be okay for the 'conservative" category, but for road engines hotter than this, the usual practice is to start with the Champion H-9 Commercial (or its equivalent in another make). You can then run a road test to check your plugs, using a procedure much like testing for fuel mixture. Accelerate hard up to high speed, declutch, and shut the motor off; if the plug base is sooty and wet, you need a hotter type, and if glazed and burned, a colder plug. Fig. 8-17 will help you here. If you're souping for extreme competition conditions, the Champ J-6 or 3 will probably be best (see accompanying tables).

Now a word about that important item of spark gap. This is critical because the voltage at which the spark "lets go" or jumps increases both with gap and compression pressure. Using large stock gaps (0.025-0.030 in.) with 8:1 or more compression increases the "punch" at the electrodes to the point where the plug will fail in a few-hundred miles. A smaller gap is the answer here — but there's a catch: At low speeds when gas turbulence in the cylinder is at a minimum, combustion lags with very short gaps, which of course, hurts our low — RPM acceleration and even causes irregular firing at idling. So we must consider low-speed performance in our spark setting.

Here are our recommendations: For the "conservative" and "medium" road engines, use a plug equivalent to the Champ H-9 Commercial and set

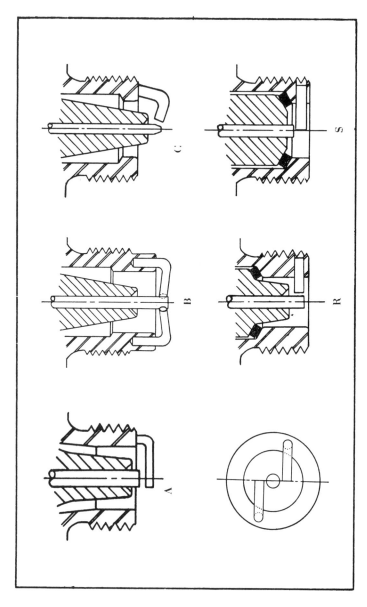

Fig. 8-18. Champion plug electrode types for automotive engines.

the gap about 0.020 in. for normal road use. For the hottest road engines, H-9 plugs should still do, but gap might need to go to 0.018 in. to keep them in; for "drag" races from a dead stop, 0.025-0.030 in. is greatly superior, but the plugs won't take it for long. For track competition where you don't need to worry about low-speed output, your plugs will stay longer and won't foul so easily if you use gaps of 0.014-0.018 in., especially with very high compression. The J-6 or J-3 heat range plugs are in order here.

IGNITION RECOMMENDATIONS

That about takes care of the basic ignition layout. Now let's go back over the problem and make a few definite recommendations:

Conservative road engine — Stock ignition should be okay here, though if your engine is a late post-1949, something should be done like converting to double breakers or a high-output coil. Plugs should be stock or in the Champ H-9 Comm. heat range and gap can be 0.020 in.

Medium road engine — We definitely need something special here. If you don't insist on winding it up over 5000 rpm in the gears, you might get away with only a Mallory or D.S.M. coil (with the old double-breaker distributor), but for best all-around results, we'd recommend a converted dual-coil system. Plugs in the H-9 range and 0.020-in. gap.

Hot road engine — A dual breaker-coil setup is essential here. Plugs should be in the Champ H-9 or J-6 heat range and gap should be 0.018 in. for durability — 0.030 in. for acceleration.

Competition engines — For most track applications with the V8, a dual ignition will be more than adequate. However, if you're loaded with dough and shooting the works, a fixed-advance magneto may be more reliable under extreme conditions and actual performance may even be a hair sharper. Whatever ignition you use, with the slower-burning racing fuels like alcohol, high-octane gas, etc., your maximum spark advance may need to be a bit more than provided by stock mechanisms in order to get maximum HP. You will need to determine this by experiment, but better figure on 30-35° with flat heads, and with the lower cylinder gas turbulence encountered with overhead valves, you may need over 40° advance. Incidentally, plugs in this category will range around the Champ J-3 or J-6 and, since you're not after back-breaking low-speed acceleration, 0.014-0.018 in. gap will keep the plugs in longer.

So that's that on the ignition problem. As you can see, there is more to it than meets the eye, and you can well understand now why it's got to be RIGHT — or our engine is a flop. Now let's skip off the beaten souping path for a bit and have a look at another strange and wonderful way of boosting that HP that has hardly been touched by hot-rodders — supercharging.

So away into the land of blowers and blown gaskets, melted pistons, stripped V-belts — and 1½ hp per cu. in.!

Champion Plug Tables

Thread Size	Heat Range	Plug Type	Hex. Size	Reach	Constr.	Stand. Gap	Electr. Type	Chief Applications
18mm	Hot ↕ Cold	R-3	1	½	2-piece	.015-.018	R	Offenhauser, Miller, Riley, Bugatti engines.
		R-7	1	½	2-piece	.019-.022	R	Mercedes, Offenhauser, Miller, Riley engines.
		R-1	1	½	2-piece	.015-.018	R	Riley, Alfa-Romeo engines; Gray, Lycoming, Elto engines.
		R-11	1	½	2-piece	.015-.018	R	Offenhauser, Miller engines; Gray, Lycoming inboard engines.
		R-11S	1	½	2-piece	.011-.014	R	Outboard racing.
		R-2	1	½	2-piece	.015-.018	R	Miscellaneous racing engines.
		R-2S	1	½	2-piece	.011-.014	R	Outboard racing.
18mm	Hot ↕ Cold	R-15	1	¾	2-piece	.015-.018	R	Alfa-Romeo and miscellaneous engines.
		R-16	1	¾	2-piece	.015-.018	R	Alfa-Romeo and miscellaneous engines.
		R-17	1	¾	2-piece	.015-.018	R	Novi, Thorne engines.
		R-18	1	¾	2-piece	.015-.018	R	Novi engines.
		RJ-19S	1	¾	1-piece	.015-.018	S	Masserati, Alfa-Romeo engines.

Fig. 8-19. Champion plug tables.

Thread Size	Heat Range	Plug Type	Hex. Size	Reach	Constr.	Stand. Gap	Electr. Type	Chief Applications
7/8"-18	Hot ↑ ↓ Cold	3 Com.	15/16	3/4	1-piece	.025	A	Kerosene, Distillate tractors, oil field engs., older trucks.
		2 Com.L	15/16	3/4	2-piece	.030	B	(same as 3 Com.)
		22	15/16	5/8	2-piece	.025	A	Buda engines; miscellaneous gas engines.
		C-4	1 1/16	5/8	2-piece	.025	C	Ford equip. mdls. A & B; other cars thru 1933.
		6	7/8	5/8	1-piece	.025	C	Willys-Knight; Buicks thru '28; old mdls. Reo, Nash.
		1 Com.	15/16	5/8	1-piece	.025	A	John Deere & miscell. tractors; Autocar & miscell. trucks.
		0 Com.	15/16	5/8	1-piece	.025	A	I.H.C. equip.; miscellaneous trucks and tractors.
18mm	Hot ↑ ↓ Cold	9 Com.	1	5/8	1-piece	.025	A	Miscellaneous tractors and industrial machinery.
		C-15	7/8	1/2	1-piece	.025	C	Buicks 1929 thru 1937; for tractors fouling 15-A.
		C-7	1	1/2	2-piece	.025	C	Miscellaneous passenger cars thru 1933.
		8 Com.	1	1/2	1-piece	.025	A	Autocar, Brockway, Diamond T trucks; B & S and miscell. eng.
		15-A	7/8	1/2	1-piece	.025	A	I.H.C. equip.; LeRoi and miscellaneous engines.
		7	1	1/2	1-piece	.025	A	Ford V-8 equip. 1934-'37; G.M. cars 1932-'36; miscell. outboards.
		13	1	1/2	1-piece	.020	R	Hall Scott & miscell. engines; miscell. passenger cars.
		6 Com.	1	1/2	1-piece	.025	A	Brockway, Corbitt, Reo, White, Federal & miscellaneous trucks.
		H-17-A	7/8	1/2	1-piece	.025	A	I.H.C. power units & farm machinery; miscellaneous engines.
		5 Com.	1	1/2	1-piece	.025	A	I.H.C. trucks; miscell. trucks severe service conditions.
		H-16-A	7/8	1/2	1-piece	.025	A	Hall Scott engines and miscellaneous applications.
14mm	Hot ↑ ↓ Cold	J-14	13/16	3/8	1-piece	.037	A	For Buicks and others fouling J-12.
		J-12	13/16	3/8	1-piece	.037	A	Buick after 1937.
		J-11	13/16	3/8	1-piece	.037	A	Chrysler Corp. cars 1932-'36.
		J-8	13/16	3/8	1-piece	.025-.028	A	Chrysler Corp. cars after '36; Chevrolet 1937-'40.
		H-10	13/16	7/16	1-piece	.025-.028	A	Ford, Lincoln, Mercury after '37; most H. C. aluminum heads.
		J-7	13/16	3/8	1-piece	.025-.028	A	Kaisers, Frazers; Nash after 1942; Willys, Studebaker.
		H-9 Com.	13/16	7/16	1-piece	.025-.028	A	Ford Commercial Cars after 1937; Miscellaneous engines.
		J-6	13/16	3/8	1-piece	.025-.028	A	Graham 1940-'41; miscellaneous commercial cars and buses.
		J-3	13/16	3/8	1-piece	.025-.028	R	Outboard engines; miscellaneous racing engines.
		J-2	13/16	3/8	1-piece	.015-.018	R	Miscell. trucks; Mack fire equip.; Gray marine & racing.
10mm	Hot ↕ Cold	Y-8	5/8	1/4	1-piece	.037-.040	A	For cars fouling Y-6; Chevrolet cars after 1940.
		Y-6	5/8	1/4	1-piece	.037-.040	A	Chevrolet trucks after 1940; Packards 1937-'42.
		Y-4-A	5/8	1/4	1-piece	.028-.030	A	Packard equipment after 1942; Cadillacs, LaSalles.

SPARK PLUG COMPARISON CHART

HEAT RANGE	THREAD SIZE	BLUE CROWN HUSKY	CHAMPION	A/C	AUTO-LITE
HOT	10MM.	T-8	Y-8	108	—
↑		T-6	Y-6	106, M-8	P-6
↓ COLD		T-4	Y-4, Y-4A, Y-5	103-S, 104 103	P-4
HOT	14MM.	M-11	J-4, J-12, J-14	47, 47-Com. 48, 49	A-11
↑		M-9	J-5, *J-5J J-11, *J-11J	46, 46-Com. *46M	A-9, AL-9 AT-8
		M-7	H-10	45, 45L-Com. 45L 45S, 45Com., *46LM	A-7, AN-7 AL-7, AH-8
		M-5, *M-5S	J-7, *J-7J, J-8 J-9, *J-9J, *J-8J	44, 44-Com. *45M	AL-5 A-5, AN-5
		M-4 Com.	H-9 Com.	43L, 43L-Com., *44M	AH-4, AT-4
		M-3 Com.	*J-10Com.J, J-10Com. J-10, J-3, -6, *J-6J	43, 43-Com. *44LM	—
↓ COLD		RM-2	JA-11, J-2	42 Com., *43M	A-3
HOT	18MM.	H-189	DL-8C, 49	18-A	BT-15
↑		H-187	D-8	14	—
		88-S	9 Com.	88, 88L-Com.	B-11, BT-10
		87-S	C-7, 8, H-8, C-15	87, 87Com., *87M	B-9
		86 86-S	8 Com., JI-15 15-A	86S, 86-Com. 86, 87S, *86M	BH-8, BT-8
		85 Com. 85-S	7, 7 Com., 13	85 Com. *85M 85S-Com.	B-7
		84 Com.	6 Com.-62, 6M 6 Com., H-17A	84, *84M	B-5
		83 Com.	5MJ, 5 Com. 5M, H-16A	83S, 83 Com. 82 Com., 82S-Com.	BH-4, BT-4
		R-82	R-1, R-3, R-7	81S-Com., 82, *83M	B-3
		R-81	R-2, R-11A, R11, R11S	81, *82M	—
		R-80	R-2A, R-2S	—	—
↓ COLD		R-80A	—	—	—
HOT	⅞″	H-179	45	19	—
↑		H-177	44	18	TT-15
		H-176	46	17	—
		†78 Com.-L 78 Com.	†3Com., †2Com.-L 22, 3, 3X	†77L-Com., †78L-Com. 78, 78S, 77	T-11 TH-10, TT-10
		76 Com.	6, C-4	76S, 76, 76-Com.	T-9, TH-8, TT-8
		75 Com.	1 Com.	75S, 75, 75-Com.	T-7
		74 Com.	0 Com.	74, 74-Com.	T-5
↓ COLD		73 Com.	00 Com., 73	73S, 73, 73-Com.	TH-4, TT-4
HOT ↑ ↓ COLD	½″ Pipe	T-F	A-25, 30, 31 29	26, 28 24	F-11

NOTE: †LONG REACH—Check piston and valve clearance unless type is recommended. Damage to piston and/or valves can result from the use of these plugs in certain engines which are designed with minimum clearance for short threads.
*"Special plugs for Marine Engines."

Fig. 8-20. Range comparisons on plug brands most popular for souped V8 engines.

CHAPTER 9

SUPERCHARGERS

MORE THAN fourteen years ago, Wild Bill Lawrence, famous stock car record man, drove a little '37 Ford across the continent in 57½ hours at a record average of 50.9 mph!

This hot little V8 was not quite 100% stock, though. Mounted on the engine was the newly-introduced McCulloch-Ford supercharger, a beautiful $125 accessory that gave Lawrence an easy 124 hp, terrific acceleration, and a top of 95 mph!

Thus did the idea of "supercharging" burst upon the stock souping scene — and fittingly enough, the V8 was the first to get the business. It's a pity that we didn't follow up the idea. Since the introduction of the McCulloch blower in 1937, there have been only five or six other installations available for the V8, with more in the offing.

WHY BOTHER?

We say this is a pity because supercharging in general is a beautiful and practical souping method. Theoretically there is no limit to the amount of HP you can pull by just putting more and more pressure to the intake manifold. In practice, of course, things don't work out quite like this and there are a hundred things that hold us back. But the fact remains that we are working with something (pressure) that has no top limit — ingenuity, engineering, and money can give us about any HP we want from it!

We can have no better proof of this fact than to look at the supercharged racing engine. From the time of the first blown race car to the present time, specific outputs — that is, HP per cu. in. — have risen steadily from 0.9 hp/cu. in. to 4½ hp. And they are still going up! Right now we're fooling around with 0.9 hp/cu. in. with our souped V8's; yet there is no reason in the world why we couldn't eventually go to 1½ hp/cu. in. or more with supercharging.

What we are getting at is this: The science of souping stock engines has reached such a high stage of development that only very drastic measures are going to pull another flock of horses. As you all know, they're already taking to those measures in the form of "super fuels." They are not depending on porting and super cams any more to get oxygen into the cylinders to burn the fuel — the boys are carrying their oxygen in a fuel can and burning an EXPLOSIVE in the engine! This can certainly lead to no good.

Why not turn instead to that other "unlimited" souping method — the supercharger? Development in this direction would be much more practical, would give us HP we could use on the road, would be safer to work with, and we might even learn something in the bargain! So we will go on record right here to urge serious speed tuners who are in it up to their necks to take a crack at supercharging. **Here's** a practically untouched field that promises rare dividends in the form of 300-400 hp Mercs and 150-mph road speeds.

We have been beating the drum a long time for this — and the advent of super fuels has made us feel even more strongly about it. Super-charging — not super fuels — is the answer to the HP problem. Let's give it a whirl. It's not a simple thing — but work with it — **experiment** with it — live with it — sweat with it — give it half a chance. It will work! Let's look it over:

SUPERCHARGING PRINCIPLES

In the first place, we covered the general subject of supercharging stock engines in some detail in our book "Souping the Stock Engine," so we won't go into it deeply here. Let's just look at basic principles and see what is available for the V8.

Back in Chapter 4, we stated a basic aim in our souping is to increase the weight of fuel-air mixture inducted into the cylinder on the intake stroke. This enables us to burn a greater weight of fuel in a given time and increases HP accordingly. Now obviously, if we can maintain our mixture at a positive pressure in the manifold, we get a proportionally greater weight of it into the cylinders than could be sucked in under atmospheric pressure ($14\frac{1}{2}$ lbs./sq. in.) and we get more HP at all **RPM**.

This is what a supercharger does. It is nothing more than a large-capacity fluid pump designed to maintain a positive pressure in the intake manifold at all speeds (with full throttle, of course). But there's a catch: There are two distinct types of superchargers available for the Ford-Merc V8 today — and thereon hangs a very important tale! It's all in the way they compress the mixture. There is (1) the "centrifugal" type, such as the McCulloch and Frenzel, which compresses by virtue of fluid momentum, and (2) the positive-displacement "Roots" type, such as the Italmeccanica, which compresses by simply pumping the fluid. Let's look them over:

THE CENTRIFUGAL TYPE

Fig. 9-1 shows the general layout of this type. It consists merely of a rotor or "impeller" disk with a number of radial blades on one side rotating in a casing. The fuel-air mixture is scooped in at the hub and whirled by the radial blades; centrifugal force pulls it toward the outside edge of the disk, where it is hurled off into the outlet casing at high velocity. From here, it literally piles up in the intake manifold — in other words, the momentum energy of the high-velocity gas is converted into pressure energy when the gas loses speed in the manifold. In the case of automobile superchargers, we turn the impeller 20,000-30,000 rpm at peak speed, which gives us an exit gas velocity off the impeller of something around 800 ft. per second; depending on our blower design and layout, this will result in a final positive pressure of about 3-6 lbs./sq. in. (of course, we could get much higher pressures by using a larger impeller such as is used in aircraft engines).

Fig. 9-1. General layout of a centrifugal-type supercharger.

Now we know that the momentum energy of any given mass increases as the SQUARE of its velocity. That's a basic physical principle. But how does this effect our centrifugal or "fluid momentum" supercharger? Obviously our pressure output here should fall off as the square of the impeller RPM — which is exactly what it does! In fact, it is even a bit worse than this because, with this type of compression, the heat generated by the work being done on the fluid is retained almost entirely within the fluid itself — which further increases the overall compression and causes the pressure output to fall off a little faster than the square of RPM. This is very bad for a road engine, as we will see later.

THE ROOTS TYPE

This type is named for its inventor, and the general layout is shown in Fig. 9-2. This blower compresses the fluid in an entirely different way than the centrifugal type. In this case, the gas is caught up on the inlet side by the interlocking rotors, carried around between rotors and casing, and simply discharged into the manifold. There is no compression within the blower at all — it just pumps gas faster than the engine can burn it and maintains a pressure just by virtue of its piling up fluid in the manifold until a state of equilibrium is reached!

From this, it would seem that our pressure output here would be constant at all speeds, but this is not quite true. In order to eliminate noise, friction, and wear in the Roots-type blower, it is constructed to run with

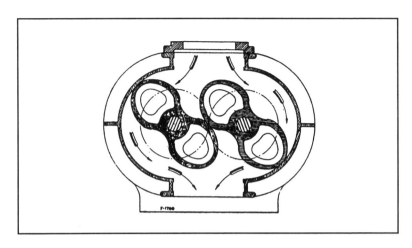

Fig. 9-2. Sectional layout of a Roots-type blower (two-lobe); arrows indicate the air flow path.

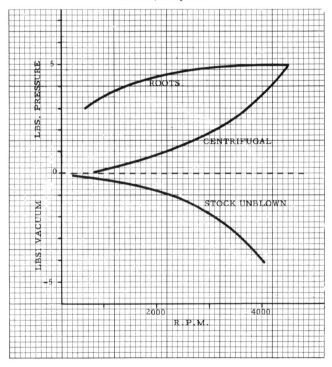

Fig. 9-3. Approximate full-throttle manifold pressure curves, with and without supercharger and assuming both blowers are designed to pump 5 lbs. boost at 4500 rpm.

a small clearance of about 0.005 in. at all points. Thus there is always a "slip loss" or leakage of gas back past the rotors, the amount depending only on the boost pressure (and not RPM). As a result, since this leakage forms a larger proportion of the total discharge at low speeds, we find that the "volumetric efficiency" of a Roots blower falls off from around 85% at peak speed (4000-8000 rpm in usual sizes) to maybe 65% in the neighborhood of 1000 rpm of the engine. This is not bad performance at all — and is certainly far superior to the centrifugal from this standpoint.

Fig. 9-3 shows approximately the full-throttle manifold pressure curves we would get with the two types, if both layouts are designed to give a boost of 5 lbs. at 4500 rpm.

CHOOSING A TYPE

Besides this all-important factor of pressure output, there are several other performance items that should influence our decision on whether to buy a centrifugal or Roots type. Here is a listing of the important ones:

Mixture heating — Compression generates heat, as we know. Compressing our fuel-air mixture in a supercharger raises its temperature, and this aggravates the engine cooling problem as well as increasing our fuel octane requirements. The centrifugal blower converts part of the heat generated by its compression into pressure, while the Roots does not compress internally at all — so none of its compression heat becomes pressure energy. (That's not a very good explanation, but it is too complicated for the time being!) For example, if both types compressed pure air from room temperature (70° F.) to 5 lbs./sq. in., the final temperature would be about 140° with the centrifugal, and 250° on the Roots. As a result, things like engine cooling, fuel octane, and spark timing will be somewhat more critical with the Roots type.

Blower power loss — For the same reasons outlined above — that is, compression heat rise, the Roots type will also require more power to pump a given flow rate because no compression heat is going into useful pressure energy. This will cut our available HP at the flywheel slightly on a given fuel octane limitation.

Fuel consumption — Any supercharger imposes a whipping motion on the ingoing fuel-air mixture, which gives much more even distribution between cylinders. Thus all cylinders can run at a fairly efficient mixture ratio instead of the wide 50% variations we often find in unblown engines — and we therefore burn less fuel for a given HP. The centrifugal type is better in this respect than the Roots, not only because it requires less power to drive, but because of the very high impeller tip speed.

Reliability — Here the Roots has it over the centrifugal. Since the centrifugal is geared about 6:1 with the crank, violent engine acceleration or deceleration such as we get when we "horse around" accelerates the blower at six times that rate, and imposes terrific stresses on the gears and bearings. This tends to cause faster wear and increases the possibility of mechanical trouble and breakdown. On the other hand, the Roots type runs at much slower speeds (nearer crank speed) and can be built very rugged all the way through. Actually, modern supercharger design is so well advanced that breakdown is no longer much of a problem in either case.

So there is the scoresheet on our two blower types. Fig. 9-4 shows how the different performance features effect our final HP curve.

Now how about slecting a type for our different souping purposes? From the above facts, it's simple. If you are designing a fast road car for long-distance highway travel, with good mileage, very high cruising speed, and plenty of "dig" between 50 and 90 mph, the centrifugal is your answer. If you are planning a hot street job that won't be used

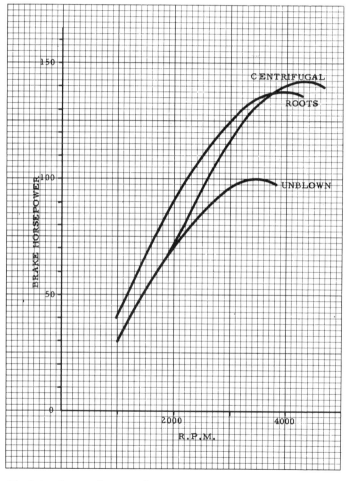

Fig. 9-4. Curves showing the approximate performance we would get with the two blower types on the pressure curves of Fig. 9-3.

primarily for long high-speed runs, that will have back-breaking acceleration at ALL speeds, the centrifugal is useless — and the Roots is your answer. For any type of road racing, the Roots should always be used; for track racing in general, the centrifugal is best, but due to the excessive dirt in the air on most tracks, supercharging is not too popular here.

Now let's take a quick look at the supercharging equipment available for the Ford-Merc V8 block:

McCULLOCH

This outfit was built by the McCulloch Engineering Co. of Milwaukee

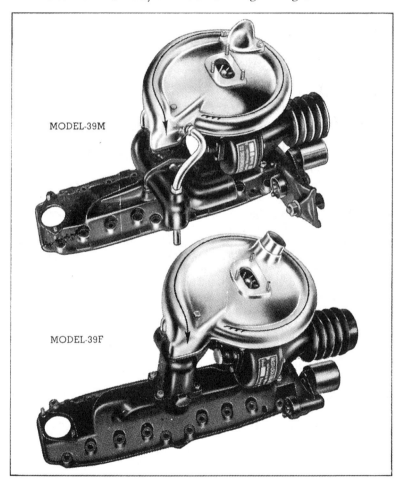

Fig. 9-5. Pioneers in the V8 supercharging field! The old McCulloch blowers for the pre-war block (Mercury at the top); both models had water jackets around the impeller housing to vaporize the fuel.

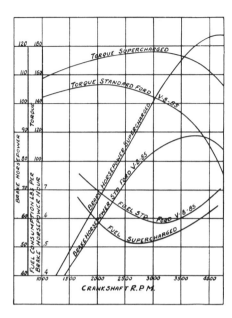

Fig. 9-6. Manufacturer's performance curves for the V8-85 with and without the McCulloch blower.

from 1937 through '40, and sold for $124.50 complete with dual exhaust system. Fig. 9-5 shows the equipment. The whole outfit mounts right on the block in place of the intake manifold and is driven by three V-belts from the crank pulley, with the generator bracketed over on one head. The blower is driven through bevel gearing in a box under the casing. Peak boost is about 4-5 lbs. and Fig. 9-6 shows performance curves of the pre-war Ford engine with and without the blower; note the substantial drop in fuel consumption. With the "39M" model, the early Mercury engine was said to develop 132 hp at 4600 rpm. (Incidentally, the HP curve in Fig. 9-6 somewhat exaggerates the blown power at low RPM; the curves should nearly coincide below 2000 rpm.)

One weakness of the old McCulloch supercharger was in the rather delicate bevel gearing which didn't stand up too well to violent acceleration. This should be kept in mind when using one. At any rate, the equipment was quite popular in its day, especially with truck owners, and there are still a few units floating around in pretty good shape. Prices usually run from $50 to $100. A good one at less than $75 would be a fair investment for a mild highway job. But if the price is over $100, you could do better by going higher and getting a new blower, such as —

FRENZEL

This is another centrifugal supercharger for the Ford-Merc block, brought out in 1950 by the Frenzel Engineering Co. of Denver. Figs. 9-7, 8, and 9 show the layout. Obviously the Frenzel boys have profited by the weaknesses of the old McCulloch, for they have set the impeller

Fig. 9-7. The new Frenzel centrifugal-type supercharger for the V8, mounting two stock Ford carbs. This outfit will pump nearly 150 hp without other changes!

Fig. 9-8. Parts photo of the Frenzel blower showing the well-shaped spiral gas channel, impeller wheel, and the very rugged gearing.

Fig. 9-9. The Frenzel supercharger mounted in the car. True top speed is 100 mph, 0-60 in 11 seconds!

in a vertical position and drive it directly through a rugged straight-spur gearing with 1-in. gear faces. This drive layout has ample strength to handle acceleration inertia loads at the 5.77:1 overall drive ratio with two V-belts. Moving parts run on five ball bearings, lubricated under engine oil pressure — all of which adds up to durability and reliability. It is a beautifully designed piece of machinery.

With two Ford carbs, the 7½-in. impeller pumps 6 lbs. boost at 4200 rpm and dynamometer tests at 3750 rpm show 136 hp for the stock Ford engine with blower, compared with 97 hp without (the actual peak would be around 145 hp at 4500 rpm). Tests at El Mirage Dry Lake in Southern California on a stock '47 Ford coupe gave the following: Without blower — top speed 84.6 mph, 0-60 mph acceleration in 16.3 seconds; with blower — top 100.05 mph, 0.60 mph acceleration in 11.0 seconds.

BESASIE

The Besasie Engineering Co. of Milwaukee has been working for several years with exhaust-driven turbo superchargers (centrifugal) for various stock car engines. They now have developed a unit for the V8 (Fig. 9-10). With this setup — that is, with the exhaust gases driving a turbine hooked directly to the blower wheel — our full-throttle boost pres-

Fig. 9-10. Besasie exhaust-turbo supercharger for the V8, mounting a Carter Olds "Rocket" carb.

sure drops off much more gradually with RPM than with a gear-driven impeller, due to the increased breathing through the valves at lower speeds. This results in improved all-around road performance.

At the same time, the exhaust "back-pressure" builds up to 3 or 4 lbs., so no baffle muffler can be used (though the gas energy loss in the turbo unit reduces noise sufficiently). Overheating might possibly be a problem on the V8. We don't know much about this Besasie model, so we won't pass judgment. Theoretically it looks terrific, especially in view of its fine ball-bearing layout and oil mist lubrication from the engine. Boost runs 6 lbs. and the peak would be around 140 hp (no curves are available).

ITALMECCANICA

This is the well-known "I.T." blower, formerly manufactured in Turin, Italy for a number of U.S. and European stock engines. The American units are now being produced in New Jersey to standard American thread sizes, etc., to simplify the maintenance problem. The Ford-Merc V8 unit is shown in Fig. 9-11.

The I.T. blowers are positive-displacement pumps of the Roots type, which give a more or less constant boost pressure and provide a terrific performance punch over the full RPM range. Fig. 9-12 shows power curves for the I.T. blower on the stock Ford, running with full accessories; with

Fig. 9-11. Italmeccanica (I.T.) Roots-type supercharger installation for the Ford-Merc. This one gives well over 150 hp and a claimed top speed of 112 mph without other changes.

engine stripped, the peak is said to be 176 hp! The boost of 6-8 lbs. provided is just about as high as we can go on pump gas and, even then, you may have to work with plugs, gaskets, spark advance, etc. with the I.T. blower.

From the standpoint of design, this unit is beautifully planned, with a rugged belt drive and good bearing layout. Another unusual and commendable feature is that there is no lubrication connection with the engine; the rear bearings run on hub grease from grease nipples, with a small reservoir containing SAE No. 90 oil feeding the front bearings (this is checked every 1000 miles). The whole I.T. layout is very neat and reliable, and gives excellent overall performance.

SPEEDOMOTIVE

Here is another Roots-type blower, produced by Speedomotive of Covina, Calif. Their Ford-Merc road unit consists of a Model 3-71 General Motors truck blower reworked and driven by four narrow V-belts. Fig. 9-13 shows the general installation mounted on a Cadillac engine (no Ford photo was available). The only dynamometer figures available are for the stock 239-cu.in. Ford block with $7\frac{1}{2}$: 1 heads and $\frac{3}{4}$ cam; with two carbs, the peak output with the blower is 195 hp at 4800 rpm at 7

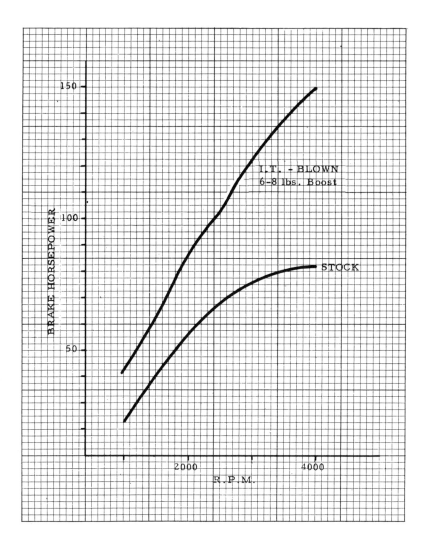

Fig. 9-12. Actual dynamometer tests of the I.T. (Italmeccanica) Roots-type supercharger installation on the 59A Ford block; both tests were run with full accessories (muffler, water pumps, etc.).

Fig. 9-13. Speedomotive Roots-type supercharger installation mounted on a Cadillac block (no Ford photo available). This layout makes use of a reworked G.M. truck blower and pumps 7½ lbs. boost pressure.

Fig. 9-14. Blower installation on Barney Navarro's famous Lakes roadster that clocked 147 mph. This is a reworked G.M. blower with four Stromberg 48 carbs. In its ultimate stage, the engine developed 240 hp from 176 cu. in. on 16 lbs. boost!

lbs. boost. This is extremely good performance, and the highly-advanced design of the G.M. blower unit promises a reliable setup.

Speaking of the G.M. 3-71 blower unit and what you can do with V8 supercharging in general, we might briefly mention Barney Navarro's interesting experiments in this souping field for his "Lakes" speed record car. Using the 239-cu. in. Ford 59A block, Barney souped as follows: Ported and relieved, 8.8:1 aluminum heads, JE pistons, Winfield SU-1 (super) cam, Spalding dual ignition. He reworked the blower with four Stromberg No. 48 carbs (0.063 jets) and drove it with four conventional automotive V-belts. Fig. 9-14 shows the layout. The peak output on gasoline was 237 hp at 5400 rpm on 10 lbs. boost — and the speed was 139 mph at El Mirage with a roadster body (1948).

From here, Barney decided to see what he could do in the 183-cu.in. SCTA Class A. This necessitated shortening the stroke to 3 in. (176 cu. in.). First experiments with metallizing and a 3/4-in. destroke on the stock Ford crank resulted in a broken shaft, though it did run a year at the Lakes before it went out.

The next step was a special 180° crank built to the 3-in. stroke with a new Winfield SU-1A billet cam to take care of the altered firing order. On its only dynamometer test, the intake valve tappets were dented (two were punctured) by the '49 Ford valves, due to the high RPM and fast cam rate. Despite this loss of timing, a peak of 218 hp at 6000 rpm on 16 lbs. boost was recorded, burning pure methanol. In later Lakes tests,. the performance was estimated as 240 hp at 6500 rpm, and the speed achieved was 146.9 mph. Not bad for a 176-cu.in. engine!

So that is about all on supercharging the V8. As we said before, here is a scandalous souping step at our fingertips. Sure, costs are high — and the problems are many — but the HP dividends just seem to keep coming as you raise that manifold pressure. We would like to see a lot more work in this field. The 300-hp road Merc may be just around the corner, and this is the way we can get it.

So now we have finished the souping job. We have gone all the way from the oil pump to the carb after our horses. But we are not quite through using our heads yet. Before we stick that ball of fire into a chassis and go out for the record — let's take pencil and paper and do a little figuring.

A consistent performer at drag races near Los Angeles, this neat little chopped coupe, with large Mercury engine, can turn 120 mph in ¼ mile.

The Miller's Crankshaft Special, an immaculate track roadster, will make 140 mph on straight run.

A successful ¾ racing car, the Dietrich Spl. employs a 274 cu. in. Merc engine.

CHAPTER 10

WHAT'LL SHE DO?

WE DON'T imagine Leadfoot Louie would even bother to read this chapter. He BUILT his engine — nobody needs to tell him what she'll do!

We are writing this chapter because we can't forget Louie's first experience with power curves and gear ratios. He had stripped the fenders off his '40 Ford coupe and dropped a dual manifold and 8:1 heads on the well-worn engine. Now he was ready to calculate the axle ratio needed for top speed. Here were his estimated factors: Peak power, 180 hp at 6000 rpm; estimated top speed, 120 mph! What happened? Not what you would expect. Louie couldn't miss a guess like a man — he had to make a mistake in his multiplication and come up with a 7.12:1 ratio! Despairing of ever finding that one in a catalog, he finally hit the road running a 4.55 rear end in SECOND GEAR.

Well, needless to say, our Louie's first brush with gear ratios ended up in a lapfull of pistons! So maybe we should take heed and pay more attention to those vital HP and RPM figures, lest we fall into similar ways. We find that amateur tuners have largely neglected this subject of true power and speed values on their souped engines. A lot of them don't know any more about a power curve than Gypsy Rose Lee, and their ideas on engine performance leave us shuddering at the thought of their being responsible for setting up a fast chassis. As a matter of fact, some close idea of the true power curve of our souped engine will be invaluable to us in setting up a chassis properly — for selecting gear ratios, clutch capacities, tire sizes, weight distributions, braking requirements, etc. So let's take some time out here with pencil and paper and estimate the performance of our souped V8.

In the first place, this is not going to be simple. Einstein himself couldn't whip out a slide rule, take a handful of figures like displacement, compression ratio, valve area, piston weight, carb size, timing, etc., and calculate your exact peak HP and RPM. It can't be done. There are just too many variable and unpredictable factors to give us that easy way out.

We must rely instead on data averaged from actual dynamometer tests on souped V8 engines. Fortunately quite a few of these tests have been run, and we have a fair volume of data available to correlate. Even then it will not be simple to estimate a complete power curve, but we have now come up with a method we think is as easy as possible under the circumstances; it will require your locating only two points on regular graph paper, drawing a straight line, and then sketching in a smooth curve. Here's how:

LAYOUT FUNDAMENTALS

We learned quite a bit about the power curve in Chapter 3 — how it was shaped and why. Now when we analyze a HP-RPM curve mathematically (which we won't discuss), we find an interesting fact that will help us a lot right now, and that is this: If we draw a straight line from the origin of the graph — that is, the zero point on both the HP and RPM scales — and if we draw this line tangent to the curve, the point where it touches is the PEAK TORQUE point. Fig. 10-1 shows what we mean here.

Now obviously we can just as well put this simple little fact to work for us in reverse to draw the HP curve. We merely locate the points of peak HP and peak torque, draw the straight line from the origin through the peak torque point, and sketch in the curve with a smooth peak at the top and tangent to the straight line. Our curve around the peak and below the torque point will be pretty much a guess, but you can get a good idea of the shaping here by studying published power curves (there are plenty of them in this book).

So that's how we will draw the curve. The next job — and this is the tough one — is to estimate those peak HP and torque points.

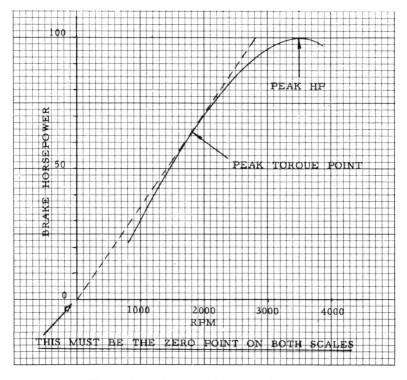

Fig. 10-1. How to determine the peak torque point on a power curve.

ESTIMATING PEAK H.P.

There are several ways we might do this. In "Souping the Stock Engine" we used two separate methods because we were working with widely different engine types. With just the basic V8, however, we believe we can simplify things, and be just as accurate, by using only a variation of the "factor" method on the HP estimation, and the "cubic inch" method on the RPM.

In the first method, we will allow for the HP increase given by each individual souping step (dual manifolds, heads, cams, etc.) by simply assigning a definite HP increase or "increment" to each step. We then just add all the increments together and add this sum, in turn, to the stock peak HP to get our new peak. For example, if our increments for an 85-hp Ford block are say 12, 17, and 8 — then we get the final increment thus: $12 + 17 + 8 = 37$, and our peak HP is $37 + 85 = 122$ hp.

So, using all available dynamometer test data, we have prepared the following increment table for the Ford-Merc V8 on gasoline:

TABLE I

Aluminum flat heads	10 (7½:1) — 20 (10:1) (See Fig. 6-4 in Chapter 6)
Milled stock heads	See Fig. 6-4 in Chapter 6
MANIFOLDS	
Regular dual	10
Super dual	13
Triple	17
Boring and stroking	Increment equal to percent increase in displacement (as 12% increase would have an increment of 12)
Porting and relieving	5-7
CAMS	
Semi	10
Three-quarter	15
Full-race	20
Super-race	25
Mushroom or roller	30
Large intake valves	5
Overhead valves	20-40 (depending on porting and general design)

That takes care of peak HP. Now for the peak RPM. For estimating this for the V8, the best method appears to be on a basis of piston displacement. Other factors equal, peak RPM will decrease somewhat as the cylinder capacity is enlarged by boring and stroking — and since the other major factor in RPM is valve timing — we can pretty well represent peak RPM on a graph as a function of the cam and the displacement. We prepared Fig. 10-2 from all available data for estimating peak RPM on our souped V8 (this graph should not be used for other basic engines).

Now let's take an example and see what we can do with the above estimation system. Suppose we have a pre-war V8-85 and we do a mild

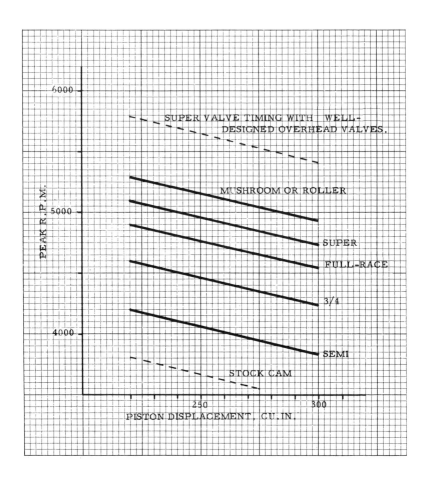

Fig. 10-2. Chat for estimating peak RPM on the Ford-Merc V8 (not applicable to other engines).

souping job, including milling the heads 0.100 in. to 7½:1 compression, Super dual manifold, and 1/8-in. bore. What is the estimated peak? For peak HP, refer to Table I and get the following increments: Heads, 5 (from Fig. 6-4 in Chap. 6); manifold, 13; with 1/8th bore, our displacement is 239 cu.in., or an increase of 239/221 = 8%, therefore our increment for this is 8. Adding these together gives us a final increment of 5 + 13 + 8 = 26 hp, and our estimated peak HP is 85 + 26 = 111 hp. Using Fig. 10-3 for estimating peak RPM, and allowing 1000 rpm for the higher compression and dual manifold, we get a peak RPM of 3800. Therefore our estimated peak for this engine is 111 hp at 3800 rpm.

ESTIMATING PEAK TORQUE

This next business may be a little tougher to savvy, but it's going to be just as easy to do and we can get pretty good accuracy on it. We learned about torque back in Chapter 3, where we went into some detail on the torque curve. You will recall that the peak of the torque curve comes at around ½ the peak RPM and that the peak torque was some 25-30% above the torque at peak HP. Now when we soup our stock engine and add hot cams, etc., we juggle these relationships somewhat — but the principle is still the same.

We can take advantage of these facts to calculate the HP at peak torque without ever actually using a torque figure. Here's how we do it: If the torque remained constant at all speeds, obviously the HP would fall off in direct proportion to RPM. However, since the torque is rising as we reduce speed, we must multiply the resulting HP by a certain ratio to get the true peak torque point on our graph. Use this formula:

$$HP_t = \frac{RPM_t \times R \times HP_p}{RPM_p}$$

where "HP_t" is the HP at peak torque; "HP_p" is the peak HP; "RPM_t" is the RPM at peak torque; "RPM_p" is the peak RPM; and "R" is the torque ratio.

Below we have listed approximate torque ratios and peak torque speeds for the different cam grinds on the V8:

Semi . . . . 1.25 @ 2100 rpm
Three-quarter . . . . 1.21 @ 2500 rpm
Full-race . . . . 1.18 @ 2900 rpm
Super-race . . . . 1.17 @ 3200 rpm
Mushroom, roller . . . . 1.10 @ 4000 rpm

Now let's calculate an example. Suppose we have already estimated our peak power as 165 hp at 4400 rpm, and we are running a full-race cam. What is the peak torque point? From the above list, our torque ratio should be 1.18 at a speed of 2900 rpm. Therefore, using the above formula, our peak torque point is:

$$\frac{2900 \times 1.18 \times 165}{4400} = 128 \text{ hp at 2900 rpm}$$

That's all there is to it. With just this and the peak HP value, we can lay out a pretty accurate power curve for our souped V8. Let's try it:

WORKING OUT EXAMPLE

Suppose we have a "59A" Ford block and we add the following "soup": Special aluminum 7½:1 rated heads; triple manifold; full-race cam; 3/16 bore and '49 Merc crank (4-in. stroke); full port and relieve jobs; special ignition and oil pressure. The problem is to plot the power curve.

In the first place, our displacement (from Table 1, Chap. 5) is 286 cu.in., or an increase of 20%. From here, and using Table 1 above, we can set up the following souping increments: Heads (8.8:1 true ratio), 17; manifold, 17; cam, 20; bore and stroke, 20; port and relieve, 7. Therefore our final HP increment is 17 + 17 + 20 + 20 + 7 = 81 — and our

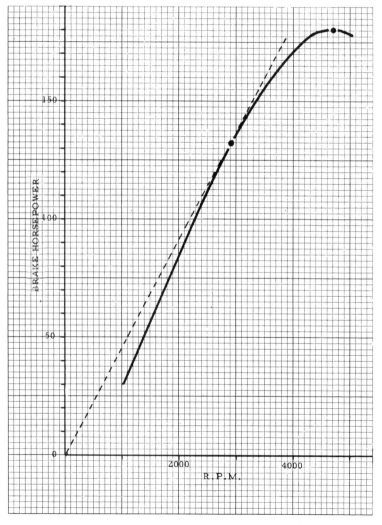

Fig. 10-3. Power curve estimation for the example in the text.

estimated peak HP is 100 + 81 = 181 hp. For estimating peak RPM, we go to Fig. 10-2, trace up on the 286- cu. in. line to the full-race cam line, and read 4700 rpm. So our estimated peak is 181 hp at 4700 rpm.

Now for the torque point. Using the previous torque ratio list, we get a ratio with the full-race cam of 1.18 at 2900 rpm; using the formula, we get our peak torque point as:

$$\frac{2900 \times 1.18 \times 181}{4700} = 132 \text{ hp at 2900 rpm}$$

The rest is simple. Fig. 10-3 shows how we construct the power curve from the above figures, according to the rules laid down earlier. Simple? We realize that this will be a good hour's job if you can't use a slide rule, but the results are well worth your while. We strongly advise that you estimate a power curve if you are setting up a fast chassis, especially a track job. (And incidentally, if you are burning pure methanol, add 10% to the peak HP figure given here.)

MORE ABOUT SOUPED PERFORMANCE

It might help you inexperienced rodders to get a sharper idea of the true HP potentialities of your V8 if we quoted a few actual figures and investigated a few actual power curves. We wouldn't bother, but we feel that it is vital to the success of your work, and to the satisfaction you get out of it, that you know what power you are getting — and what you can and can't get!

Early V8-85 blocks (pre-1939) — With the early pre-1939 Ford, displacement is 221 cu.in. and the peak is 85 hp at 3800 rpm. With maximum recommended bore and stroke of 1/8 in., displacement is 248 cu.in. and maximum potential output with flat heads on alcohol would be about 200 hp at 5200 rpm (and that's running her ragged!). We don't have many dynamometer test results on these early engines. Fig. 10-4 shows some curves and here are two Indianapolis conversions:

    1934 BOHNALITE SPECIAL — 8½:1 milled heads, dual carbs, 0.030 in. bore (223 cu.in.), reground cam; Power, 140 hp at 4400 rpm.

    1935 MILLER-FORD — 9½:1 milled heads, quad manifold, reground cam, special pistons (221 cu.in.); Power, 160 hp at 5000 rpm.

Later post-1939 Ford-Merc blocks (59A, etc.) — HP potentialities are a lot higher because we can bore out 3/16 in. and install a late 4-in. Merc crank stroked 1/8 in. to give a maximum practical displacement of 296 cu.in. At this size, the maximum at present with flat heads is about 250 hp at 5000 rpm on alcohol; with overhead valves, the top would be upwards of 300 hp at 5500 rpm.

With smaller bores and strokes, specific output — that is, HP per cu.in. — can go a little higher and peak RPM's can reach about 5200 rpm with flat heads and 5800 with overheads. Dynamometer tests on one full-

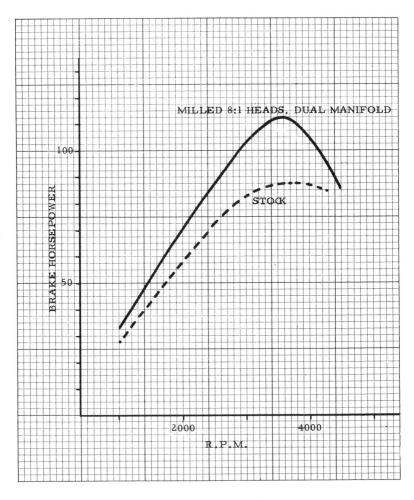

Fig. 10-4. Some dynamometer curves for the old pre-war Ford V8-85 engine (running without accessories).

race 59A block with stock bore and stroke (239 cu.in.) gave a peak of 211 hp at 4800 rpm on alcohol. That is 0.88 hp/cu.in., whereas the very large engines get only around 0.80-0.85 hp/cu.in. under these conditions because of inadequate breathing through the valves. With overhead valves, the V8 has proved itself capable of nearly 1.0 hp/cu.in. in large displacements, and it could conceivably go to 1.05 with smaller bores and strokes.

In the milder souping categories, we can do no better than to quote the figures achieved by Tom McCahill with his famous "Mechanix Illustrated Hot Rod," a stock '49 Ford V8 sedan with a Grancor-equipped engine. Here are the dynamometer figures they got (with open exhaust): In

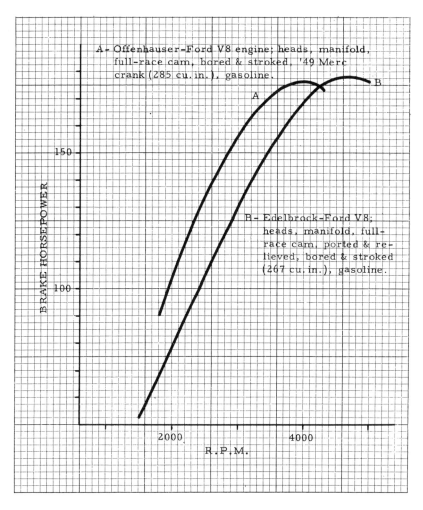

Fig. 10-5. A good example of the effect of increased piston displacement on performance; both engines use a "59A" block, full-race cam, and both develop about the same HP, but one gets its power from great displacement while the other gets it from high RPM, porting and relieving.

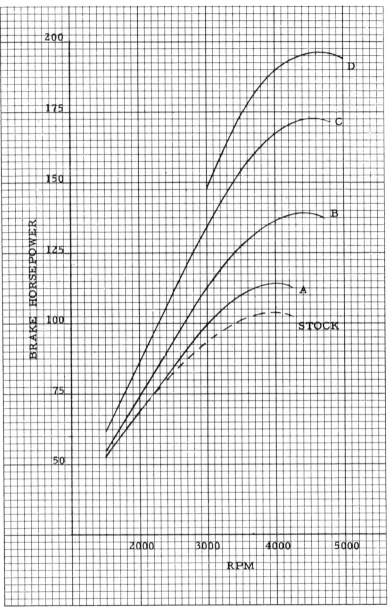

Fig. 10-6. *Typical power curves for the Edelbrock-Ford V8 conversions.*
A. Full stock, except "REGULAR" dual manifold.
B. As above, but with $8\frac{1}{2}:1$ heads and $\frac{3}{4}$ cam.
C. As above, but bored and stroked $\frac{1}{8}$ in. (267 cu. in.), ported and relieved, "SUPER" manifold, 9:1 heads, full-race cam, 80 octane pump gas.
D. As above, but super-race cam and pure methanol fuel.

stock form, the engine peaked 98 hp at 3800 rpm; by adding only Grancor 8.2:1 aluminum heads, this peak jumped to 116 hp at 3800. The engine was then torn down, bored out 0.030 in. and fitted with stock oversize Ford pistons (244 cu.in.), a dual-riser manifold mounting one large Merc carb, a reground cam about equal to a 3/4, special bearings, and a Mallory coil and condenser. This upped the peak to a reported 158 hp at 4300 rpm.

Figs. 10-5 and 6 show some HP curves for various degrees of souping on the post-1939 block. Study them carefully. Notice once again that, in all this feverish souping work, we seldom do much for HP at low RPM — simply because of the reground cam. We stick on hot heads and bore it out like a bucket — which boosts our HP at all speeds — then we have to drop in a full-race cam to get another 20 hp at the top, only to hex the whole business at 1500 rpm. If you ever rode behind a full-race mill on the road, you know what we mean; strictly "Aunt Effie stuff" at 30 mph — and even 50 — but at 70 or 80 mph, things really come to life. You don't get anything for nothing; remember that.

So we want to wind up this discussion on souped engine performance by urging you once again to "get with" the subject. Study the power curves carefully — soak in the specific output values — fix the peak figures in your mind. Pay a lot of attention to this subject, for it's as important to the overall success of a fast vehicle as choosing the right cam. Leadfoot Louie wouldn't bother with this, of course, but look at Louie now; he's screaming down the highway in his roadster at 79 mph with his full-race V8 winding 5800 — and he doesn't know whether his clutch is out or if he's got the wrong gear ratio!

Early example of the aircraft wing tank in use as streamlined dry lakes car. Engine is full-race Merc, front wheels are of motorcycle parentage. Top speed is said to have been near 150 mph (1946-47).

Beneath hood of this well-customized '41 Ford coupe rests a 270 cu. in. Mercury power plant, full-race-equipped

CHAPTER 11

THE ALMIGHTY DOLLAR AGAIN

IF YOU have five-hundred or a thousand clams in your sock, ready to shell out on a full-race V8, don't bother to read this chapter. This is for Joe Doakes down the street who has a wife and kids and a low budget — who can't put blocks together straight and whose tool collection consists of one bent screwdriver. Joe wants a hot engine in the family chariot, but he can't even afford to have the engine torn down for a reground cam.

We find there are a lot of fine chaps in similar situations, scaling up from the boy who only dreams to the man who can shoot the works. And this includes those in the middle who can spend a little, but can't do a bit of the actual work themselves for one reason or another. So let's take a moment and make some suggestions for these "forgotten men."

In the first place, in Chapter 4 we listed the various souping steps in the order of their relative HP increase per dollar spent. However, when we are severely limited in funds — and especially if we can't do the installation ourselves — that juggles things somewhat, and we have to write another set of rules. So let's start right from the bottom:

"SOUPING" FOR THE SHORT-MONEY CHAP

About the cheapest thing you can do that will have any appreciable effect on the performance of your V8, of course, is to remove the heads and have them milled 0.050-090 in. this will run about $8-12, including redoming. As we pointed out earlier, milling cast iron heads is a questionable practice — you get a very slight increase in performance and a big increase in cooling troubles and knock. You can get rid of the "ping" by retarding spark, as outlined in Chap. 8, but it costs you most of your HP increase at the peak.

Which brings us to the next step: For around $25-35, several companies sell special cast iron high-compression heads available in 7 to $8\frac{1}{2}:1$ compression. These will require a retarded spark just as milled heads, but the combustion chamber shape is better adapted to the high compression and will give a bit better performance. Don't get these if you plan to go a lot farther with your souping in the future.

From here, there is only one more direct souping move we can make without tearing the engine down — dual carbs. A word of caution: Don't invest in the cheap dual manifolds with low risers that are unevenly spaced over the ports; they have practically no effect on performance. About the least you can pay for a good dual is $40, and the extra carb will be about $12; installation cost will be quite low.

If you have gone this far with your V8, you have nearly $75 in new parts on it, probably less than $5 in labor, and about 120 hp on a late block.

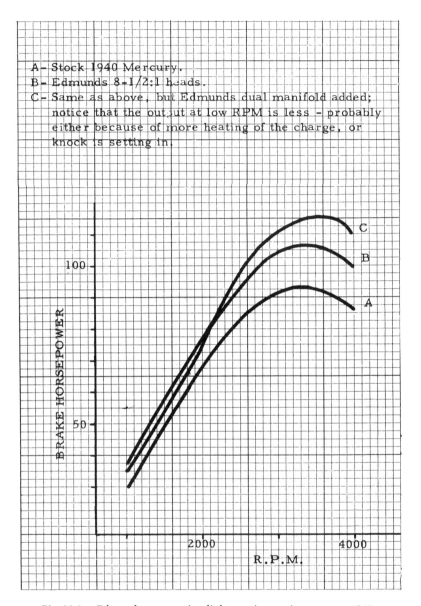

Fig. 11-1. *Edmunds curves using light souping equipment on a 1940 Merc block.*

IF YOU CAN TEAR IT DOWN

Perhaps the occasion will arise at the time when you have to have your engine torn down for a routine overhaul (bearings, rings, etc.). This is a wonderful time to put in more speed equipment. (The labor cost for just pulling and disassembling a V8 and putting it back together would be roughly $60.)

Right here we come to a vital question that few inexperienced rodders appreciate: Is it better to bore out or to install a reground cam? If you are only thinking of a performance boost at high road speeds, the cam is your answer; total price with adjustable tappets and heavy springs will be about $50, and you will get some 15 hp more at peak RPM.

But with your family bus, low-speed performance around town is apt to be pretty important — and maybe the wife won't get a kick out of having the engine idle like a cement mixer and sputter when you push it in high at 15 mph. Go back and take a look at Fig. 7-19 in Chap. 7. In other words, for low-speed performance, a reground cam is worse than nothing, while increased displacement is very desirable.

There are several companies that sell inexpensive oversize 4-ring road pistons for the V8 up to 3/16 in. oversize for $20-40 a set (these are not racing pistons, but they are good for moderate road use). The bore job will run about $20, and you will get some 8-12 hp more at the peak and a very useful increase over the full RPM range for a total investment of about $50. Another possibility here if you have the Ford engine or early Merc is to just buy a late Merc 4-in. crank and pistons for about $70, and drop them in as is for a 7-hp boost. And finally, since the above-mentioned special pistons also come in "stroker" sizes, you could combine both these steps for a total price of about $100, get a piston displacement of 276 or 286 cu.in. on the basic 239 block, and get 15-20 hp increase.

This is certainly worth thinking about. Don't just rush out for a reground cam because we listed it as the cheapest way to increase your peak HP — for if you have to have your engine torn down just to put it in, it is not as practical as aluminum $8\frac{1}{2}:1$ heads. And remember, if you are souping the car you also use for everyday transportation, that low-RPM performance is vital too, and a cam is going to play hob with this! If you still insist on the cam, you can get back some of the low-speed wallop indirectly by installing an optional 37-tooth ring gear in the rear axle (4.11:1 ratio) — but this won't help gas mileage and engine wear on the highway. On the other hand, because of the new 4-in. Merc crank, increasing piston displacement at last becomes a practical proposition for the chap with a low budget on the V8 block.

Now, we need something special in the way of ignition. Don't go overboard on this either. If you are just using your car for easy street and road travel, and are not trying to stay with every rod in town at 6000 rpm in low gear, the following steps should do with the amount of souping outlined above: If you have a pre-1949 ignition, a high-output coil at about $12 should do the trick; installation cost will be low. If you

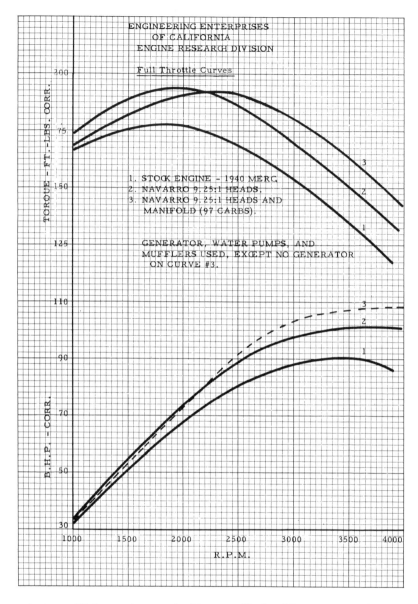

Fig. 11-2. Light souping on the '40 Merc block with Navarro equipment.

have the late engine, try a double-breaker conversion kit at $7 and use the stock coil (you can make the conversion yourself); if that sputters, go to the high-output coil.

So that about covers the problems except perhaps where to find that C-note to start with! Incidentally, if you plan to go quite a lot further than this with your souping, follow the more conventional procedures outlined earlier in the book.

CARDS ON THE TABLE

Okay — now you are going to get it! This is what WE would do if we were spending OUR OWN MONEY, and were severely limited in our expenditure all the way:

In the first case, if we couldn't ever see going beyond $30 (probably unlikely), we would have no choice but to install inexpensive cast iron 8½:1 heads and retard the spark. If we could go to around $50, but no further, we would still turn our back on dual manifolds and instead, try to dig up some second-grade aluminum 8½:1 heads, and advance spark to the mild knock point. The overall performance increase would be much better than with the dual carbs.

If we could go to $100 or a bit more, but still couldn't see pulling the engine down, we would try to squeeze in 8½:1 aluminum heads and the dual manifold (Super type with heat risers). However, if the engine needs to come down for one reason or another, and if we can see some future in our souping to maybe $300-500, we would forget all about carbs and heads for the time being. Taking advantage of the opportunity, we would sink every souping dollar in boring out 3/16 in., and installing a 4-in. Merc crank, 1/4-in. stroker 4-ring road pistons, and a 3/4 or full-race cam; if we couldn't go quite that far, we would omit the cam. This situation is a tough one to decide because of the labor expense that is more or less "wasted" in just taking the engine apart and reassembling it — so if at all possible (assuming you want to go as far as possible with your souping), you should try to finance the bore and stroke, cam, port and relieve, and oil pump work all at once while the engine is down. The other equipment can be added later (though your engine is apt to be disappointing at first without it).

At any rate, if you are limited in your funds, but do intend to eventually go as far as you can, plan very carefully. For example, don't invest in 8½:1 rated heads; when you increase your bore and stroke, the true ratio will be near 10:1, which is pretty high for road use. Just let your head work with your pocketbook instead of against it — and if you don't stray too far from the recommendations in this book, you should eventually get that top performance you are looking for!

So that's about it for this time. Before we go, here is the latest on Leadfoot Louie. He's laid up again — this time with a couple of broken ribs. He was tightening down a manifold bolt with an oversize wrench, and when Louie tightens a bolt, he doesn't mess around! His face was going from pink to purple .... the torque on the wrench was wavering around 150 lb.-ft ..... when .. WHAM! They found our hero stretched across the back of the block, the fuel pump in his stomach, multiple lacerations, and a couple of ribs out.

After that boner, we think we can almost predict how Louie will die. Picture it: He pulls off on the shoulder of a highway to fix a flat on the right front. He gets the wheel nuts off, but the wheel sticks for some reason. He's tugging for all he's worth, when .... BAM! ... the wheel flys off, Louie loses his balance, rolls over into the road — and is run over by a passing car!

Oh well, accidents will happen. But as we've said before, Louies are made, not born. This book won't help you decide whether to buy ten shares in the Brooklyn Bridge if offered, but if you're building up a hot V8, we hope it will keep you from making a Leadfoot Louie of yourself. In conclusion, we're reminded of the immortal words of Shakespeare (?):

"A man is a man who will fight with a sword,
Or tackle Mount Everest in snow;
But the bravest of all owns a '34 Ford,
Who will try for six-thousand in low!"

THE END

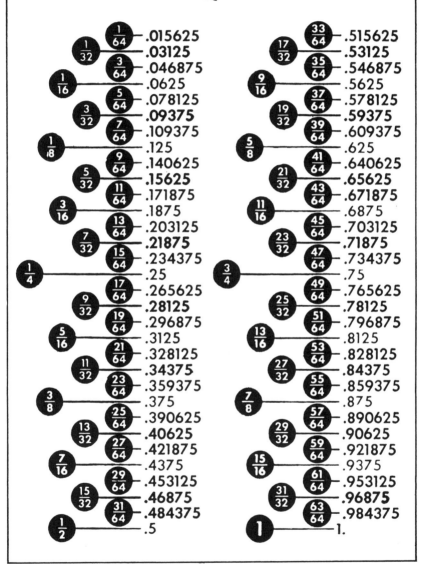

About Floyd Clymer

from the original inside cover of *Souping the Stock Engine:*

Floyd Clymer grew up with the American automobile and is a man well qualified to evoke memories of Barney Oldfield, the Stutz Bearcat, the Stanley Steamer, the Glidden Tours, the Octoauto (eight wheels), the Duck (with its back-seat steering wheel), the Tin Lizzie, and that original Horseless Carriage which mounted a life-size horse's head on the radiator. For half a century, Floyd Clymer has tested, raced and restored American automobiles. He has even invented accessories for them. Teddy Roosevelt called him the "the world's youngest automobile salesman" and he once held the motorcycle speed record to the top of Pike's Peak in Colorado. In recent years he has written and published a score of books on the automobile.

Floyd Clymer also published the 1954 *California Bill Chevrolet, GMC & Buick Speed Manual* and the 1952 *California Bill Ford Speed Manual* when Bill Fisher closed that business in the mid-50s. These two California Bill publications have also been reproduced and are again available through Fisher Books and its dealers.

Disclaimer

When this book was originally published, Floyd Clymer or Roger Huntington would never have thought to include a disclaimer. But in today's society it is important to include a disclaimer in anything that is published.

This information is republished as a fun project to make history available to auto enthusiasts. The material in this book was created around 1949. It is offered solely as an indication of how automotive performance was obtained at back then.

The parts, materials, processes or anything else shown or discussed in he book will probably not be available in the 1990s. There's no guarantee that the methods and materials presented herein will be suitable for the engine modifications or racing that you might want to do today. Over 46 years of development have changed many concepts and methods since this book was first published in 1951. The general ideas of the paths to power are still totally valid, but the specifics and recommendations may have changed in many instances.

Many of the manufacturers mentioned are no longer in business.
Most of the parts that are shown and described have not been made for many years. If you can find them they may only be available at very high prices at swap meets or through automotive enthusiasts' magazine ads, such as in Hemmings.